CIVIL ENGINEERING BRIDGE STRUCTURES

Review for the Breadth/Depth Exam in Civil Engineering

Alan Williams, Ph.D., S.E., C.Eng.
Registered Structural Engineer-California

Engineering Press
Austin, TX

Illustrations: John and Jean Foster

ISBN 1-57645-041-4

Printed in the United States of America 4 3 2 1

Engineering Press P.O. Box 200129 Austin, Texas 78720-0129

CONTENTS

The Proposed Breadth/Depth Exam

The Breadth/Depth examination format may be used as early as October 2000. At this writing the final details of the exam have not been announced, but the topic content of these exams is shown in the NCEES Appendix C table on the following page. The table shows the topic content of the exam. The morning exam will be the breadth of the topics with each topic equally weighted. The exam will be objectively scored by computer using A,B,C,D multiple choice answers. The test will probably consist of 40-60 single weight questions. The afternoon examination booklet will contain 5 exams, the candidate will choose one of the five options. In each option, a major topic consists of 65% of the option with the balance of the questions being of related topics. The afternoon exam will also be objectively scored by computer using 40-60 questions with multiple-choice answers.

Taking The Exam

Exam Dates

The National Council of Examiners for Engineering and Surveying (NCEES) prepares Civil Engineering Professional Engineer exams for use on a Friday in April and October each year. Some state boards administer the exam twice a year in their state, while others offer the exam once a year. The scheduled exam dates are:

	April	October
1999	23	29
2000	14	27
2001	20	26
2002	19	25

People seeking to take a particular exam must apply to their state board several months in advance.

List of Knowledge clusters with percentage of questions	AM Test	PM Test				
PM Test Option Number		**1**	**2**	**3**	**4**	**5**
TRANSPORTATION KNOWLEDGE	20%	65%	5%			5%
Traffic Analysis	X	X				
Facility Design Criteria (policies & guidelines)		X	X			
Construction Techniques, Equipment, and.Materials	X	X	X			X
Geometric Design (Analytical Geometry)	X	X	X			
STRUCTURAL KNOWLEDGE	20%		65%			20%
Loadings	X		X			X
Analysis	X		X			
Mechanics of Materials	X		X			
Materials	X		X			X
Member Design	X		X			X
Failure Analysis	X		X			
Design Criteria; i.e. codes			X			
WATER RESOURCES KNOWLEDGE	20%	20%	10%	65%	25%	
Hydraulics	X	X	X	X	X	
Hydrology	X	X	X	X	X	
Meteorological and Climatological Data Analysis	X	X		X	X	
Water Treatment	X			X	X	
ENVIRONMENTAL KNOWLEGDE	20%			25%	65%	
Biology (including micro- and aquatic)	X			X	X	
Bacteriology	X			X	X	
Solid/Hazardous Waste	X				X	
Ground Water and Well Fields	X			X	X	X
GEOTECHNICAL KNOWLEDGE	20%	15%	20%	10%	10%	65%
Field Exploration and Laboratory Testing	X	X	X	X	X	X
Soil Mechanics Analysis	X	X	X	X	X	X
Foundation analysis	X	X	X	X	X	X
Retaining Structures	X		X			X
Total Percentage	100%	100%	100%	100%	100%	100%

Table 1-1. State Boards of Registration for Professional Engineers

State	Mail Address	Telephone
AL	P.O. Box 304451, Montgomery 36130-4451	334-242-5568
AK	P.O. Box 110806, Juneau 99811	907-465-2540
AZ	1951 W. Camelback Rd, Suite 250, Phoenix 85015	602-255-4053
AR	P.O. Box 254 1, Little Rock 72203	501-324-9085
CA	2535 Capitol Oaks Drive, Suite 300, Sacramento 95833-2919	916-263-2230
CO	1560 Broadway, Ste. 1370, Denver 80'YO2	303-894-7788
CT	165 Capitol Ave., Rm G-3A, Hartford 06106	806-566-3290
DE	2005 Concord Pike, Wilmington 19803	302-577-6500
DC	614 H Street NW, Rm 923, Washington 20001	202-727-7833
FL	108 Hays St, Tallahassee 32301	850 521-0500
GA	166 Pryor Street, SW, Room 504 Atlanta 30303	404-656-3926
GU	P.O. Box 2950, Agana, Guam 96910	671-646-3138
HI	P.O. Box 3469, Honolulu 96801	808-586-2702
ID	600 S. Orchard, Ste. A, Boise 83705	208-334-3860
IL	320 W. Washington St, 3/FL, Springfield 62786	217-785-0877
IN	302 W. Washington St, E034, Indianapolis 46204	317-232-2980
IA	1918 S.E. Hulsizer, Ankeny 50021	515-281-5602
KS	900 Jackson, Ste 507, Topeka 66612-1214	913-296-3053
KY	160 Democrat Drive, Frankfort 40601	502-573-2680
LA	1055 St. Charles Ave, Ste 415, New Orleans 70130	504-95-85220
ME	State House, Sta. 92, Augusta 04333	207-287-3236
MD	501 St. Paul Pi, Rm 902, Baltimore 21202	410-333-6322
MA	I 00 Cambridge St, Rm 1512, Boston 02202	617-727-9956
MI	P.O. Box 30018, Lansing 48909	517-335-1669
MN	133 E. Seventh St, 3/Fl, St. Paul 55 1 01	612-296-2388
MS	P.O. Box 3, Jackson 39205	601-359-6160
MO	P.O. Box 184, Jefferson City 65102	573-751-0047
MP	P.O. Box 2078, Saipan, No. Mariana Is. 96950	670-234-5897
MT	I I I N. Jackson Arcade Bldg, Helena 59620-0513	406-444-4285
NE	P.O. Box 94751, Lincoln 68509	402-471-2021
NV	1755 E. Plumb Lane, Ste 135, Reno 89502	702-688-1231
NH	57 Regional Dr., Concord 03301	603-271-2219
NJ	P.O. Box 45015, Newark 07101	201-504-6460
NM	IO 1 0 Marquez PI, Santa Fe 87501	505-827-7561
NY	Madison Ave, Cult Educ Ctr., Albany 12230	518-474-3846
NC	3620 Six Forks Rd., Raleigh 27609	919-881-2293
ND	P.O. Box 1357, Bismarck 58502	701-258-0786
OH	77 S. High St. 16/Fl, Columbus 43266-0314	614-466-3650
OK	201 NE 27th St, Rm 120, Oklahoma City 73105	405-521-2874
OR	750 Front St, NE, Ste 240, Salem 973 1 0	503-378-4180
PA	P.O. Box 2649, Harrisburg 17105-2649	717-783-7049
PR	P.O. Box 3271, San Juan 00904	809-722-2122
RI	10 Onns St, Ste 324, Providence 02904	401-277-2565
SC	P.O. Drawer 50408, Columbia 29250	803-737-9260
SD	2040 W. Main St, Ste 304, Rapid City 57702	605-394-2510
TN	Volunteer Plaza, 3/Fl, Nashville 37243	615-741-3221
TX	P.O. Drawer 18329, Austin 78760	512-440-7723
UT	P.O. Box 45805, Salt Lake City 84145	801-530-6628
VT	109 State St., Montpelier 05609-1106	802-828-2875
VI	No. I Sub Base, Rm 205, St. Thomas 00802	809-774-3130
VA	3600 W. Broad St., Richmond 23230-4917	804-367-8514
WA	P.O. Box 9649, Olympia 98504	360753-6966
WV	608 Union Bldg., Charleston 25301	304-558-3554
WI	P.O. Box 8935, Madison 53708-8935	608-266-1397
WY	Herschler Bidg., Rm 4135E, Cheyenne 82002	307-777-6155

3

Bridge Structures

Alan Williams

Introduction

Building Structures

Bridge Structures

Foundations and Retaining Structures

Seismic Design

Hydraulics

Engineering Hydrology

Water Treatment and Distribution

Wastewater Treatment

Geotechnical Engineering

Transportation Engineering

Sample Examination Problems

Appendix A Engineering Economics

Index

HIGHWAY BRIDGE LOADS

Traffic Lanes

Since the number and location of the actual traffic lanes on a bridge deck may be changed during the life of the bridge, design lanes are used to determine the maximum possible loading condition regardless of how the bridge deck width may be demarcated for its initial intended use.

The number of design lanes is defined in AASHTO Specifications[1] in Section 3.6. For roadway widths between 20 and 24 feet, two design lanes are adopted, each equal to half the roadway width. For all other roadway widths, the design lanes are specified as being 12 feet wide and the number of design lanes is given by

$$N_L = I\!I(W/12)$$

where
$I\!I$ = integer part of the ratio
W = deck width between curbs

The location of the design lanes on the deck shall be such as to produce the maximum load effect on the member under consideration.

The determination of the number of design traffic lanes is illustrated in Fig. 3-1.

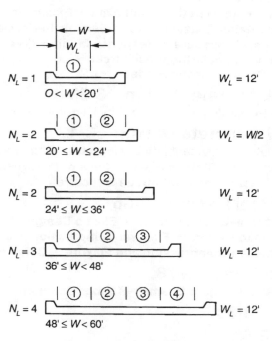

Fig. 3-1. Design traffic lanes

Vehicle Live Loads

Vehicular loading is defined in AASHTO Section 3.7. For Interstate routes, three load types are specified and these are the 72-kip HS20-44 standard truck, the equivalent lane load and the 48-kip alternative tandem loading. These are shown in Fig. 3-2. The standard truck load represents a typical heavy tractor truck with semi-trailer and is generally the critical load type for intermediate span bridges. For long-span bridges, the multiple presence of vehicles in a design lane may be more critical and the equivalent lane load represents a diverse mix of traffic. The alternative loading represents a heavy military vehicle and this may be the critical load for shorter spans.

All three types of loading are assumed to occupy a width of ten feet. The design load may be assumed to be located in any position within a design lane, without projecting beyond the lane, and may travel in either direction.

The design load adopted is that which produces the greatest load effect. When more than two design lanes are loaded, to account for the improbability of simultaneous maximum loading, the live loads are multiplied by a factor of 0.9 for three lanes and 0.75 for four or more lanes. The design lanes loaded shall be those which produce the greatest load effect.

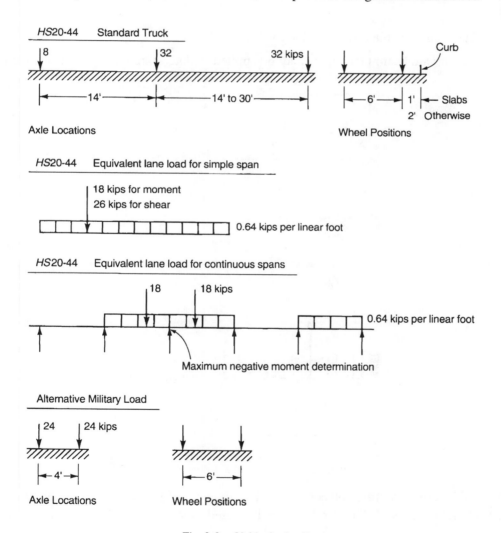

Fig. 3-2.   Vehicular loadings

Influence Lines

An influence line for a structure is a graphical representation of the variation in shear, moment, member force or external reaction due to a unit load traversing the structure. Influence line are used to determine that location of the design load on the structure which will produce the greatest load effect in a particular member. The construction of an influence line may be obtained by the application of Müller-Breslau's principle, and Maxwell's reciprocal theorem.

In accordance with Müller-Breslau's principle, the influence line for any restraint in a structure is the elastic curve produced by the corresponding unit virtual displacement applied at the point of application of the restraint. The term displacement is used in its general sense and the displacement corresponding to a moment is a rotation; that corresponding to a force is a linear deflection. The displacement is applied in the same direction as the restraint. To obtain the influence line for reaction in the prop of the propped cantilever shown in Fig. 3-3, a unit virtual displacement is applied in the line of action of R_1. The displacement produced under a unit load at any point i is δ_i. Then, applying the virtual work principle

$$R_1 \times (\delta = 1) = (W = 1) \times \delta_i$$

that is $R_1 = \delta_i$

and the elastic curve is the influence line for R_1.

The conjugate beam technique[2] may be applied to determine the actual influence line

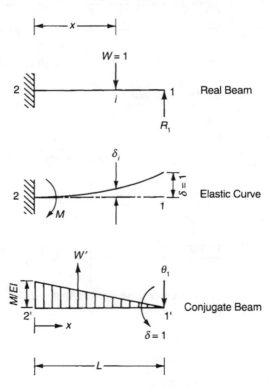

Fig. 3-3.   Müller Breslau's Principle

ordinates. The unit upward displacement applied to end 1 of the real beam produces a moment M at the fixed-end 2. This produces the indicated elastic load on the conjugate beam of

$$W' = ML/2EI$$

Taking moments about end 1' of the conjugate beam gives

$\delta = 1 = 2\,W'L/3$

Then $M = 3EI/L^2$

The intensity of loading on the conjugate beam is

$w' = 3/L^2 - 3x/L^3$

The shear on the conjugate beam is

$Q' = \int w'\,dx = 3x/L^2 - 3x^2/2L^3$

The moment on the conjugate beam is

$M' = \int Q'\,dx = 3x^2/2L^2 - x^3/2L^3$

and this is the equation of the elastic curve of the real beam and of the influence line for R_1.

In accordance with Maxwell's reciprocal theorem, the displacement produced at any point i in a linear structure due to a force applied at another point j equals the displacement produced at j due to the same force applied at i. The displacement is measured in the line of action of the applied force and the displacement corresponding to an applied moment is a rotation and to a force is a linear deflection. Referring to Fig. 3-4, and applying the virtual work principle, the deflection at point i due to a unit load at point j is

$$\delta_{ij} = \int m_j m_i\,dx/EI$$

where m_i and m_j are the bending moments produced at any section due to a unit load applied at i and j respectively. Similarly, the deflection at j due to a unit load at i is

$\delta_{ji} = dm_i m_j\,dx/EI = \delta_{ij}$

Bridge Structures

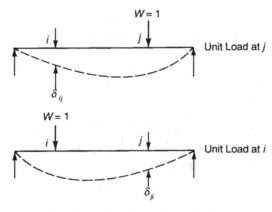

Fig. 3-4.   Maxwell's Reciprocal Theorem

Thus, the influence line for deflection at i is the elastic curve produced by a unit load applied at i.

A general procedure, therefore for obtaining the influence line for any restraint in a structure is to apply a unit force to the structure in place of and corresponding to the restraint. The elastic curve produced is the influence line for displacement corresponding to the restraint. Dividing the ordinates of this elastic curve by the displacement occurring at the point of application of the unit force gives the influence line for the required restraint. To obtain the influence line for reaction in the prop of the propped cantilever shown in Fig. 3-5, the reaction

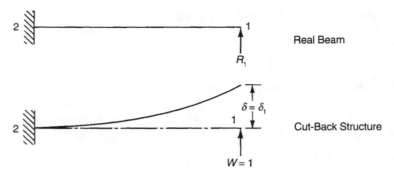

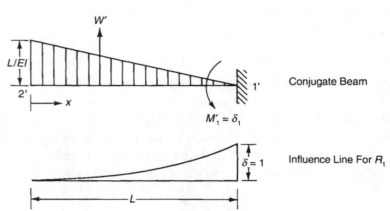

Fig. 3-5. Influence line determination

R_1 is replaced by a unit load. This produces the indicated intensity of the elastic load on the conjugate beam of

$w' = L/EI - x/EI$

The shear on the conjugate beam is

$Q' = \int w'\, dx = Lx/EI - x^2/2EI$

The moment on the conjugate beam is

$M' = \int Q'\, dx = Lx^2/2EI - x^3/6EI$

and this is the equation of the elastic curve and of the influence line for deflection of the cut-back structure at End 1. The deflection at End 1, due to the unit applied load, is given by the moment at End 1 in the conjugate beam which is

$M_1' = L^3/2EI - L^3/6EI$

$\quad = L^3/3EI$

$\quad = \delta_1$

Dividing the ordinates of the elastic curve by this displacement gives

$$M'/\delta_1 = 3x^2/2L^2 - x^3/2L^3$$

and this is the equation of the elastic curve for unit displacement at End 1 and of the influence line for R_1.

For statically determinate structures, influence lines are readily determined and examples for a simple span are shown in Fig. 3-6. The maximum bending moment, attributable to a train of wheel loads, in a simple span occurs under one of the wheels when the center of the

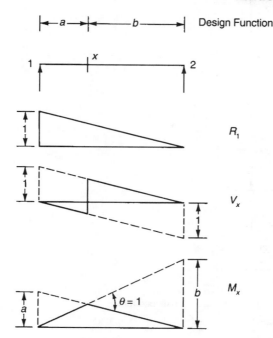

Fig. 3-6.   Simple span influence lines

span bisects the distance between this wheel and the centroid of the wheel loads. The location of the standard truck load to produce the maximum moment in a simple span exceeding 37 feet is shown in Fig. 3-7. As illustrated in Fig. 3-8, the particular load type which produces the maximum moment depends on the span length. For short spans, not exceeding 10.9 feet, a single 32-kip axle of the standard HS20-44 truck controls, and the maximum moment occurs under the axle when this is placed at the center of the span. For longer spans, not exceeding 37.12 feet, the alternative tandem load controls, and the maximum moment occurs under one axle when this is located one foot from the center of

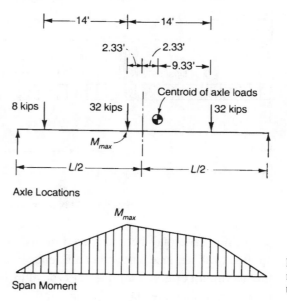

Fig. 3-7.   Maximum moment due to HS20-44 truck

the span. For longer spans, not exceeding 144.81 feet, the HS20-44 standard truck controls, and the maximum moment occurs under the central axle when this is located 2.33 feet from the center of the span. For longer spans, the equivalent lane load controls, and the maximum moment occurs at the center of the span when the 18-kip concentrated load is located there.

The maximum shear in a simple span occurs at a support and, as shown in Fig. 3-9, the particular load type which produces the maximum shear depends on the span length.

For continuous spans, the Müller-Breslau principle allows a rapid determination of the shape of the influence line and of the required location of the live load to produce the maximum effect. In accordance with AASHTO Section 3.11.4.2, only one standard truck

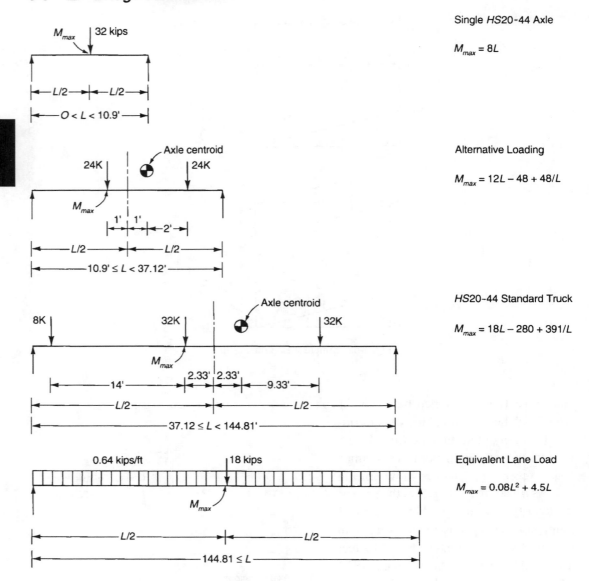

Fig. 3-8.   Maximum moments due to standard loading

per design lane may be considered on the structure. The equivalent lane load, however, is applied to all adverse parts of an influence line to give a patch loading pattern which produces the maximum effect. Examples of typical influence lines and applied loading to produce the maximum effect, are shown in Fig. 3-10.

Influence line coefficients are available[3] for a limited number of continuous spans and a limited number of span length ratios. For other situations, analytical techniques must be employed such as matrix methods[4,5] for nonprismatic structures and indeterminate trusses, moment distribution methods[6,7] for nonprismatic spans, the method of angle changes[8] for indeterminate trusses,[9] and the direct distribution of deformation for rigid frames.[10]

Example 1

For the three span bridge structure shown in Fig. 3-11, determine the maximum support reaction R_2, the maximum moment M_2 and the maximum shear V_2 due to the HS20-44 standard truck and the equivalent lane load.

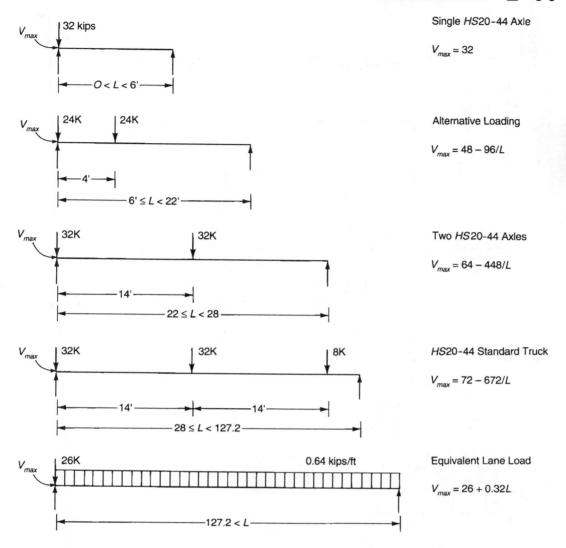

Fig. 3-9.   Maximum shear due to standard loading

Solution

The deck is statically determinate and the required influence lines are directly obtained by the application of Müller-Breslau's principle.

The influence line for R_2 is produced by introducing a unit vertical displacement at Support 2 as indicated. Placing the standard truck as shown gives the maximum reaction at Support 2 which is

$R_2 = 32(1.2 + 1.06) + 8 \times 1.032$

$\quad = 80.58$ kip

Placing the equivalent lane load as shown gives the maximum reaction at Support 2 which is

$R_2 = 26 \times 1.2 + 0.64 \times 220 \times 1.2/2$

$\quad = 115.68$ kip . . . governs.

To obtain the influence line for moment M_2 at the Support 2, the deck is cut at 2 and a unit virtual rotation imposed at 2. The elastic curve of the structure, due to this rotation is,

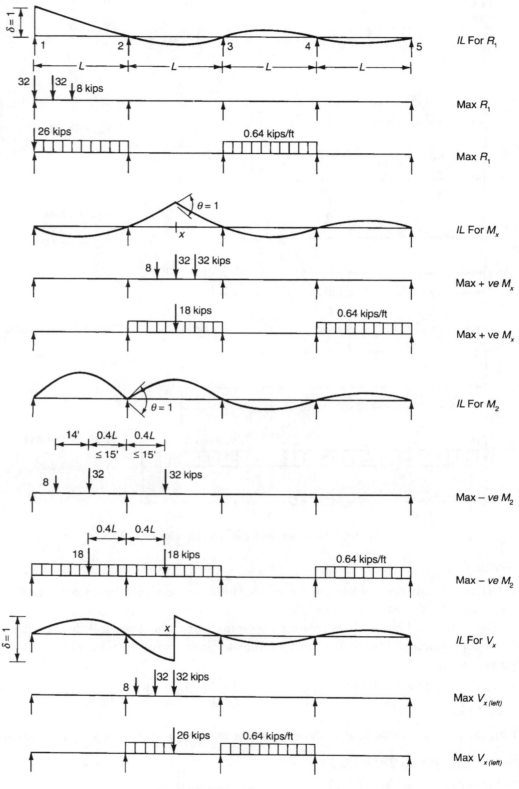

Fig. 3-10.   Maximum load effects in continuous spans

Bridge
Structures

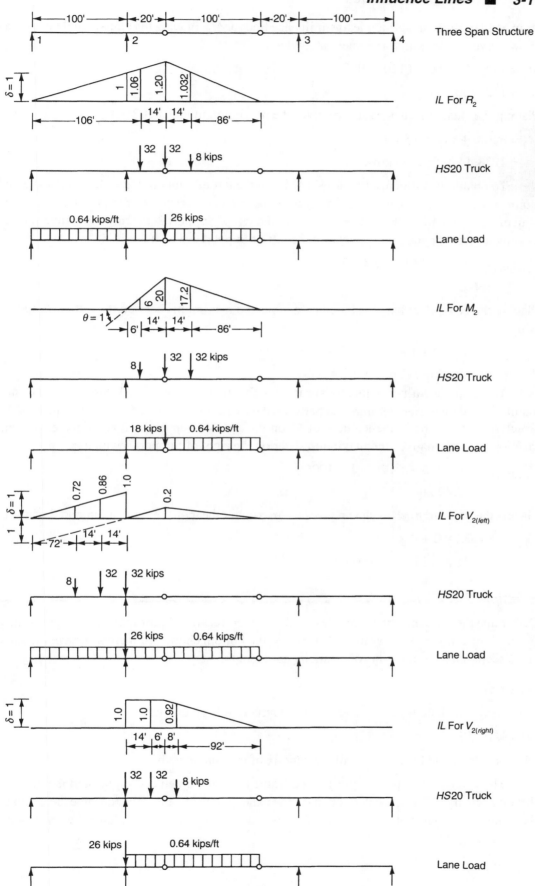

Fig. 3-11.   Details for Example 3-1

from Müller-Breslau's principle, the influence line for M_2. Placing the equivalent lane load as shown gives the maximum moment at 2 which is

$M_2 = 18 \times 20 + 0.64 \times 120 \times 20/2$

$\qquad = 1128$ kip feet.

Placing the standard truck as shown gives the maximum moment at 2 which is

$M_2 = 32(20 + 17.2) + 8 \times 6$

$\qquad = 1238.4$ kip feet . . . governs.

To obtain the influence line for shear V_2 on the left of Support 2, the deck is cut at 2 and a unit vertical displacement imposed between the two ends as shown. The elastic curve of the structure is, then, the influence line for V_2 on the left of Support 2. Placing the standard truck as shown gives the maximum shear on the left of Support 2 which is

$V_{2(LEFT)} = 32(1 + 0.86) + 8 \times 0.72$

$\qquad = 65.28$ kip

Placing the equivalent lane load as shown gives the maximum shear on the left of Support 2 which is

$V_{2(LEFT)} = 26 + 0.64(1 \times 100 + 0.2 \times 120)/2$

$\qquad = 65.68$ kip . . . governs

To obtain the influence line for shear V_2 on the right of Support 2, the deck is cut at 2 and a unit vertical displacement imposed between the two ends as shown. The elastic curve of the structure is, then, the influence line for V_2 on the right of Support 2. Placing the equivalent lane load as shown gives the maximum shear on the right of Support 2 which is

$V_{2(RIGHT)} = 26 + 0.64(1 \times 20 + 1 \times 100/2)$

$\qquad = 70.80$ kip

Placing the standard truck as shown gives the maximum shear on the right of Support 2 which is

$V_{2(RIGHT)} = 32 \times 2 + 8 \times 0.92$

$\qquad = 71.36$ kip . . . governs

Example 2

Determine the influence line ordinates for the support reaction R_1 for the two span pin-jointed frame shown in Fig. 3-12, as unit load crosses the bottom chord. All members have the same cross sectional area, modulus of elasticity, and length.

Solution

The change in slope for a member of a rigid structure is given by

$d\theta = M\ dx/EI$

and constitutes the load on a small element dx of the conjugate beam.

The change in slope of a pin-jointed frame is concentrated at the pins and is known as the angle change[8]. Thus, the deflections at the panel points of a pin-jointed frame are given by the bending moments at the corresponding points in a conjugate beam loaded with the total angle change at the panel points.

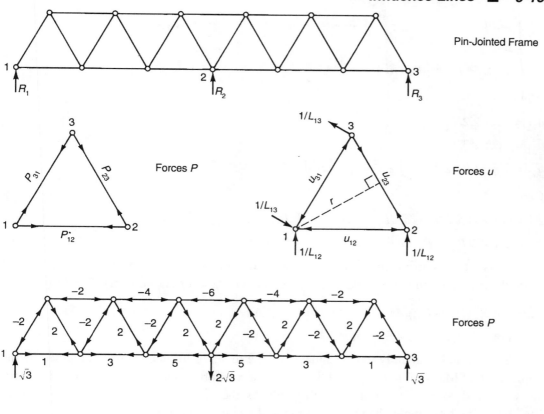

Pin-Jointed Frame

Forces P

Forces u

Forces P

Conjugate Beam:
Elastic Loads

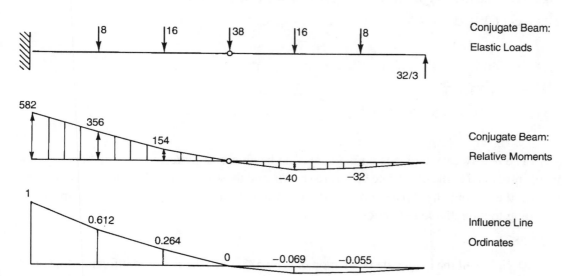

Conjugate Beam:
Relative Moments

Influence Line
Ordinates

Fig. 3-12

The tensile member forces, P, due to the applied loads, in a basic triangle of a pin-jointed frame shown in Fig. 3-12 produce positive extensions in the three members of δ_{12}, δ_{23} and δ_{31}. The angle change at 1 is given by

$$\Delta 1 = \Sigma PuL/AE$$

where u is the force in a member due to a unit couple applied to Node 1 and Node 3, as shown in Fig. 3-12, with tensile force positive. Thus

$\Delta l = \Sigma P u L / AE$

$\quad = \Sigma u \delta$

$\quad = u_{12}\delta_{12} + u_{23}\delta_{23} + u_{31}\delta_{31}$

Where

$u_{12} = -\cot \angle 2 / L_{12}$

$u_{31} = -\cot \angle 3 / L_{31}$

$u_{23} = 1/r = (\cot \angle 3 + \cot \angle 2)/L_{23}$

Hence

$\Delta 1 = \cot \angle 2 (\delta_{23}/L_{23} - \delta_{12}/L_{12}) + \cot \angle 3 (\delta_{23}/L_{23} - \delta_{13}/L_{13})$

$\quad = \cot \angle 2 (P_{23} - P_{12})/AE + \cot \angle 3 (P_{23} - P_{13})/AE$

where P is the force in a member due to the applied loads with tensile force positive and A and E are the area and modulus of elasticity, assumed constant for all members. Similarly

$\Delta 2 = \cot \angle 3 (P_{13} - P_{23})/AE + \cot \angle 1 (P_{13} - P_{12})/AE$

$\Delta 3 = \cot \angle 1 (P_{12} - P_{13})/AE + \cot \angle 2 (P_{12} - P_{23})/AE$

The frame is statically indeterminate and the required influence line is obtained by the application of Maxwell's Reciprocal theorem. The support reaction R_1 is replaced by a vertically upward force of $\sqrt{3}$ units which produces the resulting member forces shown in Fig. 3-12. The total angle changes produced at the bottom chord panel points are

$\Sigma \Delta 2 = \cot 60° [(-2 - 1 - 2 - 2) + (-2 - 2 - 2 + 2) + (2 + 2 + 2 - 3)]/AE$

$\quad = -8/(\sqrt{3} AE)$

$\Sigma \Delta 3 = \cot 60° ([-2 - 3 - 2 - 2) + (-4 - 2 - 4 + 2) + (2 + 2 + 2 - 5)]/AE$

$\quad = -16/(\sqrt{3} AE)$

$\Sigma \Delta 4 = \cot 60° [(-2 - 5 - 2 - 2) + (-6 - 2 - 6 - 2) + (-2 - 2 - 2 - 5)]/AE$

$\quad = -38/\sqrt{3} AE)$

The relative elastic loads on the conjugate beam are as indicated and the bending moments produced in the conjugate beam are shown, and these correspond to the relative deflections at the bottom chord panel points. The maximum moment in the conjugate beam, corresponding to the relative deflection at node 1 is

$M_1' = 582$

Dividing all the conjugate beam moments by 582 gives the required influence line ordinates for R_1 as shown.

Impact Allowance

In order to account for dynamic wheel load effects in the design of some bridge elements, an allowance for impact loading is added to the vehicular loads. The impact allowance is an equivalent static load, which is a function of the span length, and is expressed as a fraction of the live load to represent the dynamic effects of road surface roughness, vehicle dynamics and bridge dynamics. The impact allowance is applied to superstructures, piers and pile bents and is not required for footings, abutments, retaining walls, pedestrian loading or culverts with

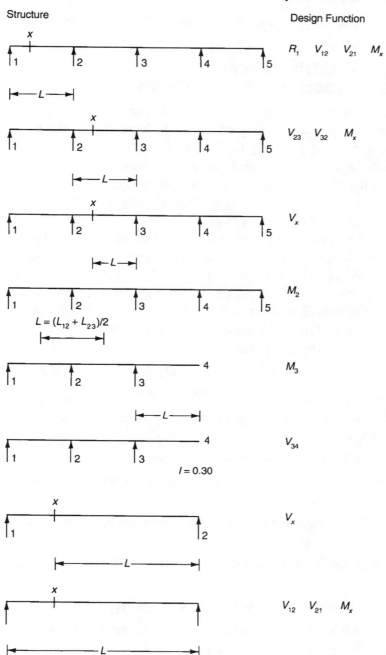

Fig. 3-13.   Impact formula length L

more than three feet of cover. A reduced impact allowance is applied to culverts with less than three feet of cover. The allowance is not applied to wood structures since these experience reduced dynamic effects due to the damping characteristics of wood and the internal friction between the structural components.

The impact fraction is given by AASHTO Formula (3–1) as

$$I = 50/(L + 125) \le 0.30$$

For simple spans, cantilevers and continuous spans, the determination of L is illustrated in Fig. 3-13.

Lateral Load Distribution and Structural Response in Bridge Decks

Dead loads are distributed to deck members in proportion to the relevant tributary areas. However, curbs, railings and wearing surfaces, which are placed after the concrete slab has cured, may be distributed uniformly to all beams.

Live loads are distributed to deck members by means of simplified empirical expressions which account for the torsional stiffness of the bridge deck system and which are simpler to apply than the more exact methods of orthotopic plate analysis,[10] grillage analysis,[11] finite strip analysis,[12] and finite element analysis.[13]

The fraction of a wheel line load which is assigned to an interior longitudinal girder, in accordance with AASHTO Section 3.23, depends upon the superstructure type, the type of girder and the girder spacing. A wheel line load is half a standard truck load or half an equivalent lane load. For widely spaced girders, wheel loads are distributed to the girders by the static, lever-rule method assuming that the flooring is hinged at each girder.

For closely spaced interior longitudinal girders, the load distribution is determined by means of the girder distribution factor

$$G = S/D$$

where, S = girder spacing in feet and D = distribution parameter.

For bridge superstructures with two or more design traffic lanes, the values of D for typical situations are:

$D = 3.75$ for $S \leq 6.5$ feet for wood plank floor on wood or steel girders.

$D = 4.25$ for $S \leq 6.5$ feet for nail laminated wood floor, not less than six inches thick, on wood or steel girders.

$D = 5.0$ for $S \leq 7.5$ feet for glued laminated wood floor, not less than six inches thick on wood girders.

$D = 4.5$ for $S \leq 7.0$ feet for glued laminated wood floor, not less than six inches thick on steel girders.

$D = 5.0$ for $S \leq 10.0$ feet for concrete floor on wood girders.

$D = 5.5$ for $S \leq 14.0$ feet for concrete floor on steel or prestressed concrete girders.

$D = 6.0$ for $S \leq 10.0$ feet for concrete Tee-beam construction.

$D = 7.0$ for $S \leq 16.0$ feet for concrete box girder.

For exterior longitudinal girders, the load distribution is determined by the lever-rule method. For the load combination of dead load, vehicular live load, impact and sidewalk live load, allowable stresses may be increased by 25 percent provided the exterior girder has a flexural capacity not less than the capacity of an interior girder. Similarly, in the case of load factor design, the beta factor may be reduced to 1.25 from the usual value of 1.67.

For shear, load distribution is determined in the same manner as for moments with the exception that, for a wheel directly over the support, the load distribution is determined by the lever-rule method.

The fraction of a wheel load which is assigned to transverse girders is determined from the girder distribution factor and values of D for typical situations are obtained from AASHTO Table 3.23.3.1:

$D = 4.0$ for $S \le 4.0$ feet for wood plank floor.

$D = 5.0$ for $S \le 5.0$ feet for nail or glued laminated wood floor, not less than six inches thick.

$D = 6.0$ for $S \le 6.0$ feet for concrete floor.

The design of a concrete slab is governed by an individual wheel load and the simplified design methods adopted in AASHTO Section 3.24 are based on Westergaard's plate theory.[14] For a slab aspect ratio not exceeding 1.5, the slab is considered supported on four sides. For a uniformly distributed load, the proportion of the load carried by the short span is given by AASHTO Equation (3-19) as

$$p = b^4/(a^4 + b^4)$$

where, a = effective length of short span,

b = effective length of long span, and

$b/a \le 1.5$

The effective span length is defined in AASHTO Section 3.24.1.2 and is illustrated in Fig. 3-14.

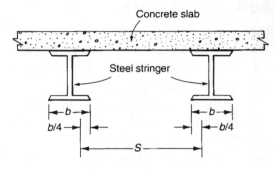

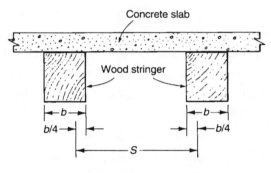

Fig. 3-14.   Effective span length, S

For a concentrated load at the center of the slab, the proportion of the load carried by the short span is given by AASHTO Equation (3-20) as

$$p = b^3/(a^3 + b^3)$$

In accordance with AASHTO Section 3.24.3.2, when main reinforcement is parallel to traffic, a wheel load is assumed distributed over a width of $E = 4 + 0.06S \le 7.0$ feet where, S is the effective span length in the relevant direction. Lane loads are distributed over a width of $E = 8 + 0.12S$.

The center half of the slab, in both directions, is designed for 100 percent of the applicable moment with the outer quarters of the slab designed for 50 percent of this value.

In the case of a simply supported slab spanning perpendicular to the traffic, the design moment for one HS20 wheel load is given by AASHTO Formula (3-15) as

$$M = 0.5(S + 2) \text{ kip feet per foot width}$$

For slabs continuous over not less than three supports, the positive and negative moments are given by

$$M = 0.4(S + 2) \text{ kip feet per foot width}$$

For a simply supported slab, not exceeding 50 feet span length, spanning in the direction of traffic, the design moment for one HS20 wheel load is given by

$$M = 0.9S \text{ kip feet per foot width}$$

For span lengths between 50 and 100 feet, the design moment is given by

$$M = (1.3S - 20) \text{ kip feet per foot width}$$

Additional Loads

The braking effects of vehicular traffic are allowed for in AASHTO Section 3.9 by applying a longitudinal force equal to five percent of the equivalent lane live load in all lanes carrying traffic headed in the same direction. The force is assumed to act horizontally at a height of six feet above the roadway surface.

Centrifugal force, due to vehicular traffic on a curved structure, is calculated as a percentage of the weight of one standard truck in each design lane since the spacing of vehicles at high speed is assumed to be large, resulting in a low density of vehicles. The applicable percentage is specified in AASHTO Section 3.10 as

$$C = 6.68S^2/R \text{ percent}$$

where S = design speed in miles per hour and R = curve radius in feet. The force is assumed to act radially, in a horizontal direction, at a height of six feet above the roadway surface.

In determining wind loads, the base design wind velocity, in accordance with AASHTO Section 3.15, is taken as 100 miles per hour. This produces a wind pressure on a beam or girder structure of

WS = 50 pounds per square foot.

Wind pressure on vehicles is represented by a horizontal force acting normal to, and six feet above, the roadway surface with a magnitude of

WL = 100 pounds per linear foot.

The vertical upward wind pressure, which must be considered when investigating overturning of the structure, is 20 pounds per square foot of deck area when wind on the live load is not included, and 6 pounds per square foot when wind on the live load is included. The force is applied at the windward quarter point of the deck width.

In accordance with AASHTO Section 3.20, lateral earth pressure shall be determined by Rankine's method with a minimum value for active pressure equivalent to that of a fluid of density 30 pounds per cubic foot. For calculating positive moments in rigid frames, one half only of the lateral pressure is assumed, and this is reflected in the values specified for the load factors in the relevant load combinations.

A surcharge pressure equivalent to two feet of fill is specified in situations when vehicular loads may approach within a horizontal distance from the top of the structure equal to one-half its height.

Other forces which must be considered include water loads, ice loads, seismic loads, thermal, creep and shrinkage forces, frictional forces and forces due to vessel collision.

Service Load Design Method

The service load, or allowable stress, design method is based on the elastic theory which assumes elastic material properties and a constant modulus of elasticity to predict the material stresses in the structure under the applied working loads. The allowable stress due to working load is defined as the yield stress or compressive strength divided by a factor of safety. This approach, however, does not ensure a constant factor of safety against failure in different structures. In particular, the margin of safety provided for an increase in the live load decreases as the ratio of live load to dead load increases.

The basic requirement for allowable stress design may be expressed in terms of stresses as

$$D + L + I \leq R$$

where

D = stress produced by dead load

L = stress produced by live load

I = stress produced by impact

R = allowable stress

Infrequently occurring overloads are accommodated by permitting a specified increase in the allowable stress and this may be expressed as

$$D + EQ \leq 1.33R$$

where EQ = stress produced by seismic effects.

Load Factor Design Method

The load factor design method, also known as the limit state or strength design method, utilizes an ultimate limit state analysis to determine the maximum load carrying capacity of the structure, and a serviceability limit state analysis to ensure satisfactory behavior at working load conditions. The objective of ultimate limit state design is to ensure a uniform level of reliability for all structures. This is achieved by multiplying the nominal, or service, loads by load factors to obtain an acceptable probability that the factored loads will be exceeded during the life of the structure. As shown in Fig. 3-15 the shaded portion of the graph denotes the probability of exceeding the factored loads.

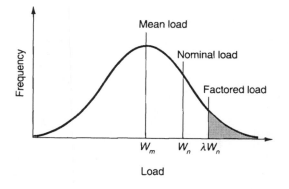 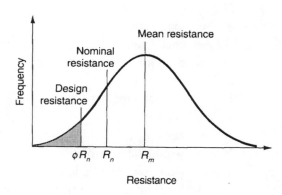

Fig. 3-15.   Probability curves for loading and resistance capacity

The factored, or ultimate, load is obtained by multiplying the nominal load by a load factor and this may be expressed as

$$W_u = \lambda W_n$$

where, λ = load factor

The load factor λ is a function of two other partial-load factors denoted by γ and β. The factor γ compensates for the unfavorable deviation of the loading from the nominal value and uncertainties in strength, methods of analysis and structural behavior. It also represents a control over the stress level developed in the structure. A value of 1.3 for γ enables the use of 77 percent of the ultimate capacity. The factor β reflects the reduced probability of the full nominal loads, in a combination of loads, being present simultaneously.

The total effect on the structure of the factored loads shall not exceed the design, or useable, resistance capacity of the structure and the basic requirement for load factor design may be expressed in terms of resistance capacity as

$$U \le \phi R_n$$

where, U = total effect on the structure of the factored load, W_u

ϕR_n = design resistance capacity of the structure

R_n = nominal resistance capacity of the structure

ϕ = strength reduction factor.

The factor ϕ is also referred to as the confidence factor, performance factor, resistance factor and capacity reduction factor. This factor allows for the possibility of adverse variations in material strength, workmanship and dimensional inaccuracies. As shown in Fig. 3-15, the shaded portion of the graph denotes the probability of exceeding the design resistance capacity of the structure.

Load Combinations

The different combinations of loads which may act on a structure are represented in AASHTO Section 3.22 by twelve groups. The loading combination for each group is given by AASHTO Equation (3-10) as

$$\text{Group (N)} = \gamma[\beta_D D + \beta_L (L + I) + \beta_C CF + \beta_E E + \beta_B B + \beta_S SF + \beta_W WS + \beta_{WL} WL$$
$$+ \beta_L LF + \beta_R (R + S + T) + \beta_{EQ} EQ + \beta_{ICE} ICE]$$

where

N	=	group number
γ	=	load factor
β	=	coefficient
D	=	dead load
L	=	live load
I	=	live load impact
E	=	earth pressure
B	=	buoyancy
WS	=	wind load on structure
WL	=	wind load on live load
LF	=	longitudinal force from live load
CF	=	centrifugal force

R = rib shortening
S = shrinkage
T = temperature
EQ = earthquake
SF = stream flow pressure
ICE = ice pressure

Table 3-1. Service Load Design Coefficients

Load	β factors for combination Group:						
	I	IA	IB	II	III	IV	VIII
D	1	1	1	1	1	1	1
$(L+I)_n$	1	2	0	0	1	1	1
$(L+I)_p$	0	0	1	0	0	0	0
CF	1	0	1	0	1	1	1
E	β_E	0	β_E	1	β_E	β_E	1
B	1	0	1	1	1	1	1
SF	1	0	1	1	1	1	1
WS	0	0	0	1	0.3	0	0
WL	0	0	0	0	1	0	0
LF	0	0	0	0	1	0	0
R + S + T	0	0	0	0	0	1	0
EQ	0	0	0	0	0	0	0
ICE	0	0	0	0	0	0	1
γFactor	1	1	1	1	1	1	1
% Basic stress	100	150	*	125	125	125	140

*As specified by operating Agency

Table 3-2. Load Factor Design Coefficients

Load	β factors for combination Group:						
	I	IA	IB	II	III	IV	VIII
D	β_D	β_D	β_D	β_D	β_D	β_D	β_D
$(L+I)_n$	1.67	2.2	0	0	1	1	1
$(L+I)_p$	0	0	1	0	0	0	0
CF	1	0	1	0	1	1	1
E	β_E	0	β_E	β_E	β_E	β_E	β_E
B	1	0	1	1	1	1	1
SF	1	0	1	1	1	1	1
WS	0	0	0	1	0.3	0	0
WL	0	0	0	0	1	0	0
LF	0	0	0	0	1	0	0
R + S + T	0	0	0	0	0	1	0
EQ	0	0	0	0	0	0	0
ICE	0	0	0	0	0	0	1
γFactor	1.3	1.3	1.3	1.3	1.3	1.3	1.3

For service load design, the applicable load factors and the permissible percentage of the basic allowable stresses are summarized in Table 3-1 where,

$(L + I)_n$ = live load plus impact for standard truck or equivalent lane loading

$(L + I)_p$ = live load plus impact for Agency abnormal load vehicle

β_E = 1.0 or 0.5 for lateral loads on rigid frames

β_E = 1.0 for vertical and lateral loads on all other structures

For load factor design, the applicable load factors are summarized in Table 3-2 where,

β_E = 1.3 for lateral earth pressure on retaining walls

β_E = 1.3 or 0.5 for lateral loads on rigid frames

β_E = 1.0 for vertical earth pressure

β_D = 1.0 for flexural and tension members

β_D = 0.75 for columns with minimum axial load and maximum moment

β_D = 1.0 for columns with maximum axial load and minimum moment.

The load combinations may be classified as:

Group I = permanent loads plus the primary vehicular loading consisting of the standard truck or equivalent lane load

Group IA = permanent loads plus the maximum design load which may be applied in an emergency.

Group IB = permanent loads plus the operating agency's abnormal load vehicle.

Group II = permanent loads plus wind load on the structure without vehicular loading

Group III = permanent loads plus primary vehicular loading plus wind load on the live load and reduced wind load on the structure.

Group IV = permanent loads plus vehicular loading plus loads originating from rib shortening, shrinkage and temperature effects

Group VIII = permanent loads plus vehicular loading plus ice pressure

Example 3

Fig. 3-16 shows the cross section of the three-span reinforced concrete Tee-beam bridge structure shown in Fig. 3-11. For HS20 loading, determine the applicable factored moment for design of the central beam at Support 2. The concrete barrier rail has a weight of 0.4 kip per linear foot.

Solution

The dead load acting on the central beam consists of the beam self weight plus the tributary portion of the deck slab plus, in accordance with AASHTO Section 3.23.2.3.1.1, one third the weight of the two barrier railings. The applicable dead load is

$W_D = 0.15(7 \times 16/12 + 10.25 \times 9/12) + 2 \times 0.4/3$

$= 2.82$ kips per linear foot.

The dead load bending moment at Support 2 is

$M_D = W_D(20 \times 10 + 50 \times 20)$

$= 3384$ kip feet

In accordance with AASHTO Section 3.6.3 the 24-foot wide roadway width is divided into two lanes each 12 feet wide. From Example 3-1, the maximum moment at Support 2 is produced by the standard truck and for a single truck, the value of the moment is

$M' = 1238.4$ kip feet

In accordance with AASHTO Section 3.8.2.2 the loaded length for determining the impact fraction is

$L = 100 + 20$

$\quad = 120$ feet

The impact fraction, as given by Formula (3-1) is

$I = 50/(L + 125)$

$\quad = 50/(120 + 125)$

$\quad = 0.204$

Thus, for a single HS20 standard truck the value of the moment at support 2 due to live load plus impact is

$M'' = M'(1 + I)$

$\quad = 1238.4(1 + 0.204)$

$\quad = 1491$ kip feet

The beam spacing exceeds 10 feet and, in accordance with AASHTO Table 3.23.1, wheel loads from the standard truck are distributed to the beams by the static, lever-rule

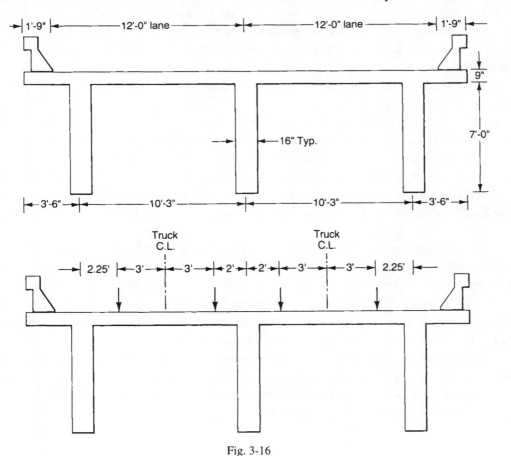

Fig. 3-16

method, assuming that the concrete deck is hinged at each beam. Maximum loading is produced in the central beam by locating one truck in each lane, as shown in Fig. 3-16, with one line of wheels a minimum distance of two feet from the lane edge, as specified in AASHTO Fig. 3.7.7A. Since only two lanes are loaded, in accordance with AASHTO Section 3.12.1, no reduction in the intensity of loading is necessary. The total moment distributed to the concrete girders is, then,

$$M_L = 2\,M'' \times 5.25/10.25$$
$$= 2 \times 1491 \times 5.25/10.25$$
$$= 1527 \text{ kip feet}$$

The factored design moment for Group I loading is given by AASHTO Table 3.22.1A as

$$M_u = \gamma[\beta_D D + \beta_L(L + I)]$$
$$= 1.3(1 \times M_D + 1.67 \times M_L)$$
$$= 1.3(3384 + 1.67 \times 1527)$$
$$= 7714 \text{ kip feet.}$$

SEISMIC DESIGN

Design Procedure

The objective of a seismic design procedure is to design a structure which may be damaged in an earthquake but which will not collapse and which can quickly be put back into service. The analysis procedure to be adopted is specified in AASHTOSD Specifications[17] Section 4.

A detailed seismic analysis is not required for single span bridges. However, minimum support widths are required, in accordance with AASHTOSD Section 3.10, and the connection of the superstructure to the substructure is designed to resist the product of the dead load reaction, the site coefficient, and the site acceleration coefficient, as specified in Section 3.11.

For multi-span regular structures with a maximum of six spans and having no abrupt changes in weight, stiffness, or geometry, the single-mode spectral method may be adopted. The method assumes a predominant single mode of vibration which allows a statical method of analysis to be utilized. The paameters which define a regular bridge ae given in AASHTOSD Table 4.2B.

The response spectrum or multimode spectral method is used for complex structures with irregular geometry. Several modes of vibration contribute to the overall response of the structure and a space frame analysis program with dynamic capabilities is required.

In order to determine the seismic response of the structure, several factors must be considered and these include the acceleration coefficient, importance classification, performance category, site coefficient and response modification factors.

The selection of the design procedure depends on the type of bridge, the magnitude of the acceleration coefficient and on the degree of acceptability of loss of operation. The single-mode spectral method is defined as Procedure 2 and the multimode spectral method is defined as Procedure 3, in AASHTOSD Section 4.1.

Acceleration Coefficient

The acceleration coefficient A, given in the contour maps in the Specification, is an estimation of the site dependent design ground acceleration expressed as a percentage of the gravity constant g. The values of A range from 4 to 80. The acceleration coefficient corresponds to ground acceleration values with a recurrence interval of 475 years which gives a ten percent probability of being exceeded in a fifty-year period. The acceleration coefficient corresponds to the effective peak acceleration[18] in bedrock, which are based on historical records and geological data, and, for sites located near an active fault zone, the coefficient may exceed the 80 percent contour value. Numerical values for A are obtained by dividing contour values by 100.

Elastic Seismic Response Coefficient

The elastic seismic response coefficient, or lateral design force coefficient C_s, is a function of the seismic zone, the fundamental period of the bridge and the site soil conditions. The value of the lateral design force coefficient is given by AASHTOSD Formula (3-1) as

$$C_s = 1.2AS/T^{2/3} \leq 2.5A$$

where
S = site coefficient or amplification factor for a specific soil profile
T = fundamental period of the bridge.

The specific site soil profile considerably influences the ground motion characteristics and three profile types and corresponding site coefficients are defined in Section 3.5.
Soil profile type I consists of rock or rock with an overlying layer of stiff soil less than 200 feet deep. The applicable value for the site coefficient is

$$S = 1$$

Soil profile type II consists of a stiff clay layer exceeding 200 feet in depth and has a site coefficient of

$$S = 1.2$$

Soil profile type III consists of a soft to medium clay layer, at least 30 feet deep, and has a site coefficient of

$$S = 1.5$$

Soil profile type IV consists of a soft clay or silt layer, exceeding 40 feet deep, and has a site coefficient of

$$S = 2.0$$

The single-mode spectral procedure assumes that the first mode of vibration predominates during the seismic response of the structure, and this is the case for regular structures. The mode shape may be represented by the elastic curve produced by the application of a uniform unit virtual load to the structure. From consideration of the kinetic and potential energies in the system, values of the fundamental period of vibration and of the generalized seismic force are obtained.[19,20] The analysis for seismic response along the longitudinal axis of the bridge may be reduced to the expression.

$$T = 2\pi\sqrt{m/k}$$

where
m = mass of the system and k = stiffness of the system.

The analysis for seismic response in the transverse direction requires the use of numerical integration techniques[21] to evaluate the relevant expressions.

The fundamental period T and the equivalent static force are obtained by using the technique detailed in Section 4.4 and involves the determination of the integrals given in the expressions (4-5), (4-6) and (4-7) where the limits of the integrals extend over the whole length of the bridge superstructure. These expressions simplify when the dead weight per unit length of the superstructure and tributary substructure is constant, and when the displacement profile is constant as is the situation in determining longitudinal effects, as shown in Fig. 3-17. Hence, for longitudinal seismic force, the expressions for α, β and γ reduce to

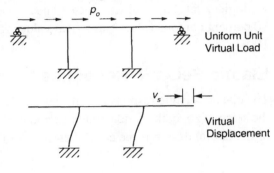

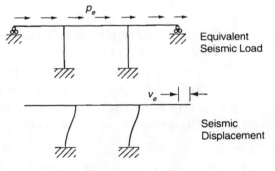

$$\alpha = \int v_s(x)\, dx$$

$$= v_s \int dx$$

$$= v_s L.$$

$$\beta = \int w(x)\, v_s(x)\, dx$$

$$= w v_s \int dx$$

$$= w v_s L$$

$$\gamma = \int w(x)\, v_s(x)^2\, dx$$

$$= w v_s^2 \int dx$$

$$= w v_s^2 L$$

Fig. 3-17.   Longinal loads and displacements

where

$v_s(x)$ = displacement profile due to p_o,

v_s = total longitudinal displacement of the structure due to p_o,

p_o = uniform unit virtual load,

$w(x)$ = distribution of dead weight per unit
length of the superstructure, and
tributary substructure = w for a constant dead weight.

The fundamental period is given by expression (4-8) as

$$T = 2\pi \sqrt{\gamma / p_o g \alpha}$$

$$= 2\pi \sqrt{w v_s / p_o g}$$

$$= 0.32 \sqrt{W v_s / P_o}$$

$$= 0.32\sqrt{W/k}$$

$$= 0.32\sqrt{\Delta_w}$$

where

$W = wL$ = total weight of superstructure and tributary substructure

$P_o = p_oL$ = total applied virtual load

k = total stiffness of the structure

Δ_w = longitudinal displacement in inches due to the total dead weight acting longitudinally

Hence, the lateral design force coefficient may be obtained from Formula (3-1) and the equivalent static seismic loading is given by expression (4-9) as

$$p_e(x) = p_e = \beta C_s w(x)\, v_s(x)/\gamma = wC_s$$

and this produces the longitudinal displacement v_e as shown in Fig. 3-17.

The total elastic seismic shear is given by

$$V = p_e L = wLC_s = WC_s$$

The procedure for determining the transverse seismic response is shown in Fig. 3-18. The abutments are assumed to be rigid and to provide a pinned end restraint at each end of the superstructure. A transverse uniform unit virtual load is applied as shown and is resisted by the lateral stiffness of the superstructure and by the stiffness of the central column bent. As shown in Fig. 3-19, the displacements produced are the sum of the displacements in the cutback structure due to the unit virtual load, plus the displacement in the cut back structure due to the reaction in the column bent. The value of the reaction in the column bent is obtained by equating the displacements of Node 2 in case (i), (ii) and (iii). The displacement of Node 2 in the original structure, as shown at (i), is

$$\delta_2 = k_C R$$

where

k_C = stiffness of the column bent

   = $12EI_C/H^3$ for a fixed ended column

R = reaction in the column bent

I_C = moment of inertia of column

H = height of column

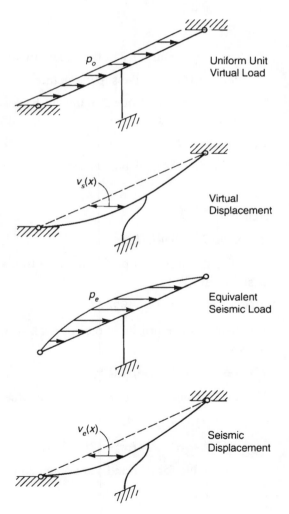

Fig. 3-18.   Transverse loads and displacements

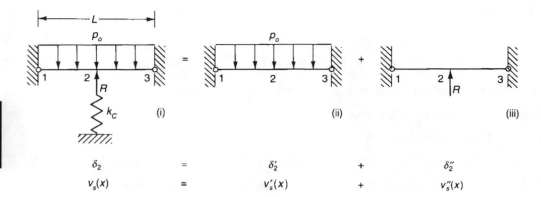

Fig. 3-19.   Transverse displacements

The displacement of Node 2 in the cut-back structure due to the uniform unit virtual load, as shown at (ii), is

$\delta_2' = 5L^4/384EI$

where

E = modulus of elasticity of the superstructure

I = lateral stiffness of the superstructure.

The displacement of Node 2 in the cut-back structure due to the reaction in the column bent, as shown at (iii), is

$\delta_2'' = -RL^3/48EI$

Equating these displacements gives

$\delta_2 = \delta_2' + \delta_2''$

Hence

$R = 5L^4/(8L^3 + 384EIk_C)$

The displacement profile in the cut-back structure due to the uniform unit virtual load, as shown at (ii), is

$v_s'(x) = x(L^3 - 2Lx^2 + x^3)/24EI$

The displacement profile in the cut-back structure due to the reaction in the column bent, as shown at (iii), is

$v_s''(x) = -Rx(3L^2 - 4x^2)/48EI$

The displacement profile in the original structure, as shown at (i), is

$v_s(x) = v_s'(x) + v_s''(x)$

$= [2x^4 + 4Rx^3 - 4Lx^2 + (2L - 3R)L^2x]/48EI$

The value for α, β and γ are obtained from expressions (4-5), (4-6) and (4-7) as

$\alpha = \int v_s(x)\, dx$

$$\beta = \int w(x)\, v_s(x)\, dx$$

$$\gamma = \int w(x)\, v_s(x)^2\, dx$$

The limits of integration extend over the whole length of the bridge and the numerical values of the integrals may be obtained by means of a calculator[21]. Alternatively, values of the integrands may be computed at discrete intervals over the length of the bridge and the numerical integration performed manually.

The fundamental period is obtained from expression (4-8) as

$$T = 2\pi\sqrt{\gamma / p_o g \alpha}$$

The seismic response coefficient is obtained from Formula (3-1) as

$$C_s = 1.2\, AS/T^{2/3}$$

The equivalent seismic loading is defined by Formula (4-9) as

$$p_e(x) = \beta C_s w(x)\, v_s(x)/\gamma$$

This equivalent static load may now be applied to the structure and the resultant forces calculated.

Importance Classification

The importance classification is defined in AASHTOSD Section 3.3 and two categories are specified. An importance classification of I is assigned to essential bridges which for social or security considerations must remain functional after an earthquake. For this situation a design is required which will ensure the continued operation of the facility. An importance classification of II is assigned to non-essential bridges.

Seismic Performance Category

The seismic performance category is a function of the acceleration coefficient and the importance classification and is defined in AASHTOSD Section 3.4. The four categories are shown in Table 3-3 and these provide flexibility in the specified requirements for selection of the design procedure, minimum support lengths, and substructure design details.

Table 3-3. Seismic Performance Category

Acceleration Coefficient	Essential Bridges (Importance Classification I)	Other Bridges (Importance Classification II)
A ≤ 0.09	A	A
0.09 < A ≤ 0.19	B	B
0.19 < A ≤ 0.29	C	C
0.29 < A	D	C

Analysis Procedure

AASTOSD Section 4.1 defines four analysis procedures. Procedure 2 is the single mode spectral analysis technique, and Procedure 3 is the multimode spectral analysis technique. The procedure selected depends on the seismic performance category and on the bridge classification, and is summarized in Table 3-4. A multi-span bridge with a uniform mass distribution, and with the stiffness of adjacent supporting members differing by not more than 25 percent, is classified as a regular structure. In this type of bridge, the fundamental mode of vibration predominates during the seismic response of the structure and higher modes of vibration do not significantly effect the distribution of seismic forces. An irregular bridge is one that does not satisfy the definition of a regular bridge and, in this type of structure, the higher modes of vibration significantly effect the seismic response.

Table 3-4. Analysis Procedure

| SPC | Bridges with 2 or more Spans | |
	Regular	Irregular
A	N/A	N/A
B	1 or 2	3
C	1 or 2	3
D	1 or 2	3

A detailed seismic analysis is not required for single span bridges or for bridges classified as seismic performance category A. However, minimum support widths are required, in accordance with AASHTOSD Sections 3.10 and 5.3, and the connection of the superstructure to the substructure is designed to resist the forces, as specified in Sections 3.11 and 5.2.

Response Modification Factors

To design a bridge structure to remain within its elastic range during a severe earthquake is uneconomical. Limited structural damage is acceptable provided that total collapse is prevented and public safety is not endangered. Any damage produced in a severe earthquake should be readily detectable, accessible and repairable. To achieve this end, the response modification factor R is specified in AASHTOSD Section 3.7, and this represents the ratio of the force in a component which would develop in a linearly elastic system to the prescribed design force. The response modification factors are selected to ensure that columns will yield during a severe earthquake, while connections and foundations will have little if any damage. The response modification values for substructure components reflect the non-linear energy dissipation capability, the increase in natural period and damping, and the ductility and redundancy of the component. Thus, the R-factor for a single column is 3 and for a multiple column bent is 5, which is an indication of the redundancy provided by the multiple column bent. The R-factor is applied to moments only. Elastic design values are adopted for axial force and shear force unless the values corresponding to plastic hinging of the columns are smaller, in which case the smaller values are used.

The R-factors of 1.0 and 0.8 assigned to connectors requires connectors to be designed for 100 percent or 125 percent of the elastic force. This is to ensure enhanced overall integrity of the structure at strategic locations with little increase in construction costs. However, the connector design forces need not exceed the values determined using the maximum probable plastic hinge moment capacities developed in the columns. The maximum probable capacity or overstrength capacity results from the actual material strength exceeding the minimum specified strength.

Table 3-5. Response Modification Factors

Substructure	R-factor
Wall Type Pier:	
Strong axis	2
Weak axis	3
Reinforced Concrete Pile Bents:	
Vertical Piles Only	3
One or More Batter Piles	2
Single Columns:	3
Steel or Composite Pile Bents:	
Vertical Piles Only	5
One or More Batter Piles	3
Multiple Column Bent:	5

Bridge Structures

Table 3-6. Response Modification Factors

Connections	R-factor	Design force
Superstructure to abutment:		
Single span	N/A	$A \times DL$
SPC A	N/A	$0.2 \times DL$
SPC B, C, D	0.8	Elastic/R
Expansion joints:	0.8	Elastic/R
Pinned columns:		
SPC A	N/A	$0.2 \times DL$
SPC B, C, D	1.0	Elastic/R
Fixed column:		
SPC A	N/A	$0.2 \times DL$
SPC B	1.0	Elastic/R
SPC C, D	N/A	Plastic hinge forces

Combination of Orthogonal Forces

AASHTOSD Section 3.9 requires the combination of orthogonal seismic forces to account for the directional uncertainty of the earthquake motion and for the possible simultaneous occurrence of earthquake motions in two perpendicular horizontal directions. The combinations specified are:

Load Case 1: 100 percent of the forces due to a seismic event in the longitudinal direction plus 30 percent of the forces due to a seismic event in the transverse direction.

Load Case 2: 100 percent of the forces due to a seismic event in the transverse direction plus 30 percent of the forces due to a seismic event in the longitudinal direction.

Load Combinations

For superstructure and substructure design down to base of columns, in accordance with AASHTOSD Section 6.2 and 7.2, the load combination Group for seismic performance categories B, C, and D, is given by

$$\text{Group Load} = \gamma(\beta_D D + \beta_B B + \beta_S SF + \beta_E E + \beta_{EQ} EQ)$$
$$= 1.0(D + B + SF + E + EQ)$$

where

D = dead load

B = buoyancy

SF = stream flow pressure

E = earth pressure

EQ = elastic seismic force for either Load Case 1 or Load Case 2 divided by the appropriate R-factor.

For service load design, a 50 percent increase is permitted in the allowable stresses for structural steel and a $33^{1}/_{3}$ percent increase is permitted for reinforced concrete.

For footings, pile caps and piles, the same Group Load is applicable and the appropriate R-factor, for seismic performance category B, is half that of the substructure element to which the footing is attached. For seismic performance catgories C and D, an R-factor of unity is applied.

Column Plastic Hinges

The determination of shear and axial forces in column bents, due to plastic hinging, is shown in Fig. 3-20 for transverse seismic loading.

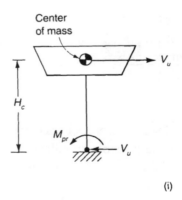

Fig. 3-20. Column plastic hinges: Transverse earthquake

For a single column, as shown at (i), the maximum probable or overstrength moment capacity at the foot of the column is

$$M_{pr} = \phi M_n$$

where

M_n = nominal moment capacity.

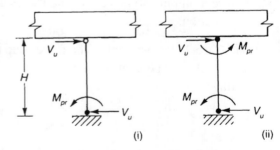

Fig. 3-21. Column plastic hinges: Longitudinal earthquake

The shear developed in the column is given by

$$V_u = M_{pr}/H_C$$

where

H_C = height of center of mass

For a double column bent with both columns fixed at top and bottom, as shown at (ii), the shears developed in the left and right columns are given by

$$V_{uL} = 2M_{prL}/H$$

$$V_{uR} = 2M_{prR}/H$$

where

H = height of column

The axial force developed in the columns is obtained by equating moments of external forces about the base of one column and is given by

$$P_{uL} = -P_{uR}$$
$$= [H_C(V_{uL} + V_{uR}) - (M_{prL} + M_{prR})] /B$$

Shear and axial forces, due to plastic hinging, for longitudinal seismic loading is shown in Fig. 3-21.

For a single column, pinned at the top, as shown at (i), the shear developed in the column is given by

$$V_u = M_{pr}/H$$

For a single column fixed top and bottom, as shown at (ii), the shear developed in the column is given by

$$V_u = 2M_{pr}/H$$

REINFORCED CONCRETE DESIGN

Strength Design Principles

The strength design procedure adopted in the AASHTO specifications is identical with the procedure adopted in the ACI[23] Building Code, and this is adequately covered in Chapter 2 of this text.

Serviceability Requirements

In order to ensure satisfactory performance under service load conditions, the distribution of reinforcement, fatigue characteristics and deflection of flexural members is controlled.

Deflections due to service live load plus impact are limited by AASHTO Section 8.9 to

$$\delta_{max} = L/800$$

where

L = span length

To provide additional comfort for pedestrians, the deflections of bridge structures which are also used by pedestrians are limited to

$$\delta_{max} = L/1000$$

To achieve these limits, AASHTO Table 8.9.2 provides expressions for the determination of minimum superstructure depths. Alternative expressions are given in the ACI Recommendations.[24]

Actual deflections may be calculated in accordance with AASHTO Section 8.13. The modulus of elasticity of concrete is given as

$$E_c = 57,000 \sqrt{f_c'}$$

and the effective moment of inertia may be taken as the moment of inertia of the gross concrete section.

When the yield strength of the reinforcement exceeds 40,000 pounds per square inch, to control flexural cracking of the concrete, the distribution of tension reinforcement shall be in accordance with AASHTO Section 8.16.8. The value of the parameter z, given by AASHTO Equation (8-61), is limited to 170 for moderate exposure conditions and 130 for severe exposure conditions where z is given by

$$z = f_s(d_c A)^{1/3}$$

where

f_s = stress in reinforcement at service load $\leq 0.6 f_y$

d_c = concrete cover measured to center of reinforcement closest to tensile face of concrete

A = effective concrete area in the tension block per equivalent number of bars

In accordance with AASHTO Section 8.17, in order to prevent sudden tensile failure of a flexural member, a minimum area of tensile reinforcement is required in order to provide a moment capacity of

$$M_{min} = 1.2 M_{cr}$$

where

$M_{cr} = f_r I_g / \bar{y}$ = cracking moment of section

$f_r = 7.5 \sqrt{f_c'}$ = modulus of rupture of normal weight concrete

I_g = moment of inertia of gross concrete section, neglecting reinforcement

$\bar{y}$ = distance of neutral axis from tension face

The minimum reinforcement ratio to satisfy this requirement is given by[24]

$$\rho_{min} = I_g f_c'(10 + I_g / \bar{y} b d^2) / \bar{y} b d^2 f_y$$

To control cracking in the side faces of members exceeding three feet in depth, longitudinal skin reinforcement is provided in the lower half of the effective depth, having a total area, in accordance with AASHTO Section 8.17.2 of

$$A_{s(min)} \geq 0.012 (d - 30) \text{ square inches/foot}$$

The spacing of this reinforcement is given by

$$s = d/b$$

$$\leq 12 \text{ inches}$$

Fatigue stress limits are governed by the magnitude of the stress in the reinforcement, the range of stress, and the shape of the deformations on the bar which act as stress raisers. The range between maximum and minimum stress levels, due to dead load and service live load plus impact, must not exceed the value given by AASHTO Equation (8-60) as

$$f_f = 21 - 0.33f_{min} + 8r/h \text{ kip per square inch}$$

where

f_{min} = the algebraic minimum stress, due to dead load and service live load plus impact, with tension positive and compression negative

Bridge Structures

r = base radius of deformations

h = height of deformations

When the actual dimensions of the deformations are not known, it is adequate to assume that the ratio

$r/h = 0.3$ and,

$$f_f = 21 - 0.33f_{min} + 2.4$$
$$= 23.4 - 0.33f_{min}$$

The actual stress in the reinforcement at service load level is determined by the straight line, elastic theory, method.

Spacing limits for reinforcement are specified in AASHTO Section 8.21. For cast-in-place concrete the clear distance between parallel bars in a layer shall be not less than

(i) $1.5d_b$

(ii) $1.5 \times$ maximum aggregate size

(iii)   1.5 inches

The minimum clear distance required between layers of reinforcement is one inch.

For concrete exposed to earth or weather, AASHTO Section 8.22 specifies a minimum concrete cover to primary reinforcement of two inches and to secondary reinforcement of 1.5 inches.

Example 4

The three span reinforced concrete Tee-beam bridge structure, shown in Fig. 3-11 and Fig. 3-16 and analyzed in Examples 3-1 and 3-3, has a concrete compressive strength of 3250 pounds per square inch and uses grade 60 reinforcement.

(i)   Determine the maximum and minimum service load moment for the central beam, at the center of span 12, for HS20 loading.

(ii)   Calculate the corresponding factored design moment.

(iii) Using an effective depth of 88 inches and number 11 bars, determine the tensile reinforcement required.

(iv)   Calculate the maximum live load deflection at the center of span 12.

(v)   Determine the minimum moment capacity required to prevent sudden tensile failure.

(vi)   Calculate the value of the reinforcement distribution parameter z.

**Bridge
Structures**

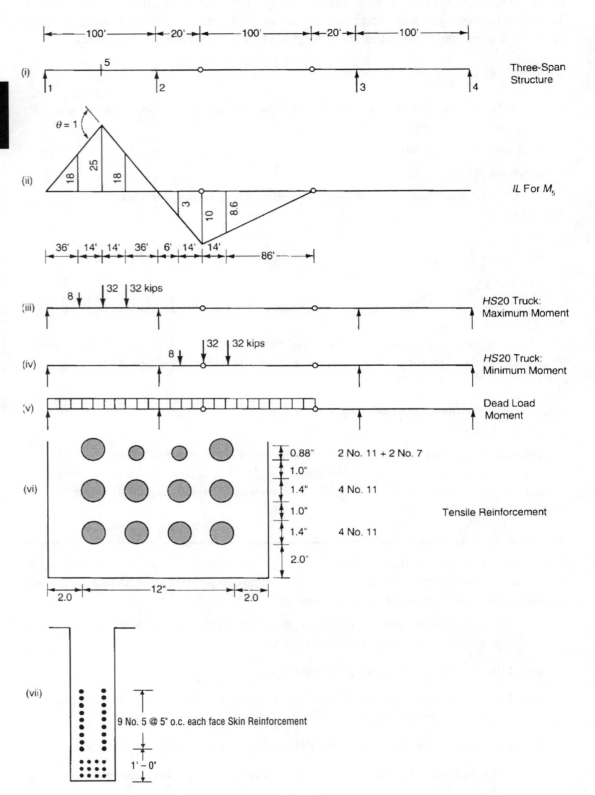

Fig. 3-22

(vii) Determine the amount of longitudinal skin reinforcement required.

(viii) Calculate the range of stress produced by service loading.

Solution

(i) To obtain the influence line for moment M_5 at the center of span 12, the deck is cut at 5 and a unit virtual rotation imposed on the two ends. The elastic curve of the structure due to this rotation is the influence line for M_5 as shown in Fig. 3-22(ii).

**Bridge
Structures**

The standard truck loading governs for both maximum and minimum moments. Placing the standard truck as shown at (iii), gives the maximum service live load at 5 which is

$M_{L(max)} = 32(25 + 18) + 8 \times 18 = 1520$ kip feet

Placing the standard truck, as shown at (iv), gives the minimum service live load at 5 which is

$M_{L(min)} = -32(10 + 8.6) - 8 \times 3 = -619$ kip feet

The dead load moment on the central beam at 5 is obtained as shown at (v) and is given by

$M_D = 2.82 \times 0.5 (25 \times 100 - 10 \times 120) = 1833$ kip feet

For the maximum live load moment, the loaded length for determining the impact fraction is obtained from AASHTO Section 3.8.2.1 as

$L = 100$ feet

The impact fraction, as given by AASHTO Formula (3-1) is

$I = 50/(L + 125)$

$= 50/(100 + 125)$

$= 0.222$

The maximum live load plus impact moment due to HS20 loading is

$M'_{L(max)} = M_{L(max)} (1 + I)$

$= 1520 (1 + 0.222)$

$= 1857$ kip feet

The total moment distributed to the central girder, as determined in Example 3-3, is

$M''_{L(max)} = 2 M'_{L(max)} \times 5.25/10.25$

$= 1902$ kip feet.

The maximum serviceability limit state moment on the central beam at 5, due to dead load and live load plus impact is

$M_{5(max)} = M_D + M''_{L(max)}$

$= 1833 + 1902$

$= 3735$ kip feet.

For the minimum live load moment, the loaded length for determining the impact fraction is

$L = 120$ feet

The impact fraction, from AASHTO Formula (3-1) is

$I = 50/(L + 125)$

$$= 50(120 + 125)$$

$$= 0.204$$

The minimum live load plus impact moment due to HS20 loading is

$$M'_{L(min)} = M_{L(min)}(1 + I)$$

$$= -619 (1 + 0.204)$$

$$= -745 \text{ kip feet}$$

The total moment distributed to the central girder is

$$M''_{L(min)} = 2 M'_{L(min)} \times 5.25/10.25$$

$$= -763 \text{ kip feet}$$

The minimum moment on the central beam at 5, due to dead load and live load plus impact is

$$M_{5(min)} = M_D + M''_{(min)}$$

$$= 1833 - 763$$

$$= 1070 \text{ kip feet}$$

(ii) The factored design moment, at Section 5 of the central beam, for Group I loading is given by AASHTO Table 3.22.1A as

$$M_u = \gamma[\beta_D D + \beta_L(L + I)]$$

$$= 1.3 (1.0 \times M_D + 1.67 \times M''_{(max)})$$

$$= 1.3 (1833 + 1.67 \times 1902)$$

$$= 6512 \text{ kip feet}$$

(iii) The effective compression flange width is given by AASHTO Section 8.10 as the minimum of

(a) $b = L/4$

$$= 100/4$$

$$= 25 \text{ feet}$$

(b) $b = b_w + 12h_f$

$$= 16 + 12 \times 9$$

$$= 124 \text{ inches}$$

(c) $b = S$

$$= 10.25 \times 12$$

$$= 123 \text{ inches} \ldots \text{ governs}$$

Assuming that the stress block lies within the flange, the required tension reinforcement is determined from the principles of AASHTO Section 8.16 as

$$A_s = 0.85bdf'_c\left[1 - \sqrt{1 - 2K/0.765f'_c}\right]/f_y$$

where

$$K = 12M_u/bd^2$$

$$= 12 \times 6512/(123 \times 88^2)$$

Hence

A_s = 16.72 square inches

Provide 10 Number 11 bars and 2 Number 7 bars, as shown at (vi) to give

A_s = 16.80 square inches

The stress block depth is given by

$a = A_s f_y / 0.85 b f_c'$

$= 16.8 \times 60 / (0.85 \times 123 \times 3.25)$

$= 3$ inches

$< h_f$ . . . satisfactory

   (iv) The recommended minimum depth of the superstructure, in accordance with AASHTO Table 8.9.2, is

$h_{min} = 0.07L$

$= 0.07 \times 100$

$= 7$ feet

   The recommended minimum depth of the superstructure, in accordance with ACI[24] Table 8.5.2, is

$h_{min} = (L + 9)/18$

$= (100 + 9)/18$

$= 6.06$ feet

The depth provided is

$h = 7.75$ feet

Table 3-7. Member Properties for Example 3-4

Part	A	y	I	Ay	Ay^2
Beams	4032	42	790,272		
Flange	2970	88.5	20,047		
Total	7002	—	—	432,189	30,374,230

   In calculating live load deflection, AASHTO Section 8.13 requires all traffic lanes to be fully loaded with the moment of inertia based on the full superstructure gross section excluding curbs and railings. The moment of inertia is obtained as shown in Table 3-7. Hence,

$\bar{y} = \Sigma Ay / \Sigma A$

$= 432,189/7002$

$= 61.7$ inches

$I_g = \Sigma I + \Sigma Ay^2 - \bar{y}^2 \Sigma A$

$= 4,508,270$ inches^4

The modulus of elasticity, for short term loads, is given by AASHTO Section 8.7 as

$E_c = 57,000 \sqrt{f_c'}$

Bridge
Structures

$$= 57,000 \sqrt{3250} / 1000$$

$$= 3250 \text{ kips per square inch}$$

The central deflection produced in Span 12 with a standard truck in each lane positioned as indicated at (iii) is given by

$$\delta_5 = 2 \times 32L^3/48EI_g + 2 \times 36 \times 12 (32 + 8)(3L^2 - 4 \times 36^2 \times 144)/48EI_g$$

$$= 2304 \times 10^6/EI_g + 2573 \times 10^6/EI_g$$

$$= 0.333 \text{ inches}$$

$$= L/3605$$

$$< L/800 \ldots \text{ satisfactory}$$

(v) The modulus of rupture of the concrete is given by AASHTO Section 8.15.2 as

$$f_r = 7.5 \sqrt{f_c'}$$

$$= 7.5 \sqrt{3250}$$

$$= 428 \text{ pounds per square inch}$$

The cracking moment of the superstructure section is

$$M_{cr} = f_r I_g / \overline{y}$$

$$= 428 \times 4,508,270/(61.7 \times 1000 \times 12)$$

$$= 2606 \text{ kip feet}$$

In accordance with AASHTO Section 8.17 the minimum required factored design moment capacity is

$$M_{min} = 1.2 M_{cr} = 1.2 \times 2606 = 3127 \text{ kip feet} < M_u \ldots \text{ satisfactory}$$

(vi) The height of the centroid of the tensile reinforcement is given by

$$\overline{c} = (4 \times 1.56 \times 2.7 + 4 \times 1.56 \times 5.1 + 2 \times 1.56 \times 7.5 + 2 \times 0.6 \times 7.24)/16.8$$

$$= 4.81 \text{ inches}$$

The total area of the concrete tension block is

$$A_T = 2 \overline{c} b_w$$

$$= 2 \times 4.81 \times 16$$

$$= 153.8 \text{ square inches}$$

The equivalent number of Number 11 bars is

$$n = A_s/1.56$$

$$= 16.8/1.56$$

$$= 10.77 \text{ bars}$$

Effective concrete area in the tension block is

$$A = A_T/n$$

$$= 153.8/10.77$$

$$= 14.28 \text{ square inches per bar}$$

The maximum serviceability limit state moment due to dead load and live load plus impact is

$$M_{5(max)} = 3735 \text{ kip feet}$$

In accordance with AASHTO Section 8.15.3 the modular ratio is taken to the nearest whole number and is given by

$$\begin{aligned} n &= E_s/E_c \\ &= 29{,}000/3250 \\ &= 9 \end{aligned}$$

Assuming the neutral axis, as determined by linear elastic theory, lies within the flange the tension reinforcement ratio is

$$\rho = A_s/bd$$
$$= 16.8/(123 \times 88)$$
$$= 0.00155$$

The elastic design parameters and the reinforcement stress may be obtained as

$$k = (n^2\rho^2 + 2n\rho)^{0.5} - n\rho$$
$$= 0.154$$

$$j = 1 - k/3$$
$$= 0.949$$

$$f_s = 12M_{5(max)}/jdA_s$$
$$= 31.954 \text{ kip per square inch}$$

The neutral axis depth is given by

$$kd = 0.154 \times 88$$
$$= 13.6 \text{ inches}$$
$$> h_f \dots \text{unsatisfactory}$$

However, an exact analysis[25] indicates a negligible discrepancy, with the steel stress

$$f_s = 31.695 \text{ kips per square inch}$$

Hence, the reinforcement distribution parameter is given by AASHTO Equation (8–61) as

$$z = f_s \sqrt[3]{d_c A}$$
$$= 31.954 \sqrt[3]{2.7 \times 14.28}$$
$$= 108 \text{ kips per square inch}$$
$$< 130 \dots \text{satisfactory}$$

In addition, the steel stress $f_s < 0.6f_y \dots$ satisfactory

(vii) The area of longitudinal skin reinforcement required in each side face of the web is given by AASHTO Section 8.17.2 as

$$A_{s(min)} \geq 0.012 \, (d - 30)$$
$$= 0.012 \, (88 - 30)$$
$$= 0.70 \text{ square inches per foot}$$

The required spacing is 12 inches maximum and providing nine number 5 bars in each face, as shown at (vii), gives

A_{sl} = 0.74 square inches per foot

> $A_{s(min)}$ . . . satisfactory

(viii) The minimum serviceability limit state moment due to dead load and live load plus impact is

$M_{5(min)}$ = 1070 kip feet

The minimum reinforcement stress is given by

$f_{min} = 12M_{5(min)}/jdA_s$

= 9.154 kips per square inch

The actual range between maximum and minimum reinforcement stress, due to dead load and service live load plus impact is

$f_f = f_{max} - f_{min}$

= 31.954 – 9.154

= 22.8 kips per square inch.

The allowable range is given by AASHTO Equation (8-60) as

$f_{f(all)}$ = 23.4 – 0.33f_{min}

= 23.4 – 0.33 × 9.154

= 20.4 kips per square inch

< f_f . . . unsatisfactory

PRESTRESSED CONCRETE DESIGN

Transfer Limit State

At the transfer limit state, the prestressing force is applied to the concrete section, and the resultant stresses are obtained as the superposition of the axial stress due to the initial prestressing force, the flexural stresses caused by the bending moment produced by the prestressing force, and the flexural stresses caused by the bending moment produced by the self weight of the member. The allowable stresses in the concrete and the prestressing steel, at transfer, are specified in AASHTO Section 9.15 and are illustrated in Fig. 3-23 where

f'_{ci} = compressive strength of concrete at transfer

≥ 3500 pounds per square inch, post-tensioned member

≥ 4000 pounds per square inch, pretensioned member

f'_s = ultimate strength of prestressing steel

f_y^* = yield stress of prestressing steel

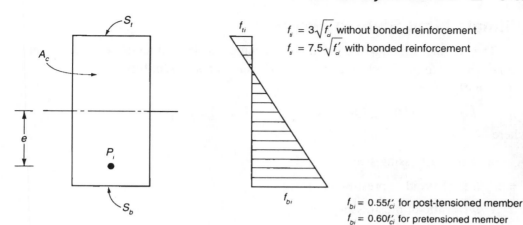

$f_{ti} = 3\sqrt{f'_{ci}}$ without bonded reinforcement

$f_{ti} = 7.5\sqrt{f'_{ci}}$ with bonded reinforcement

$f_{bi} = 0.55f'_{ci}$ for post-tensioned member

$f_{bi} = 0.60f'_{ci}$ for pretensioned member

Concrete Stresses

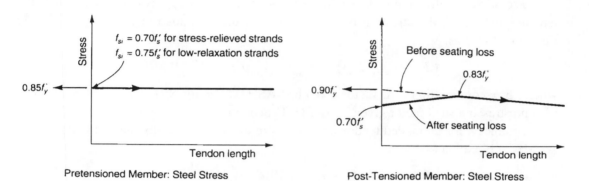

Pretensioned Member: Steel Stress

Post-Tensioned Member: Steel Stress

Fig. 3-23.   Allowable stresses at transfer

Serviceability Limit State

At the serviceability limit state, all long term prestress losses due to creep, shrinkage and relaxation have occurred to give a final prestressing force of P_e. The resultant stresses in the section are due to the final prestressing force, the self weight of the member, additional superimposed dead load, and the live load. The allowable stresses in the concrete and the prestressing steel under service loads are specified in AASHTO Section 9.15 and are illustrated in Fig. 3-24 where, f_{se} = effective steel prestress after losses.

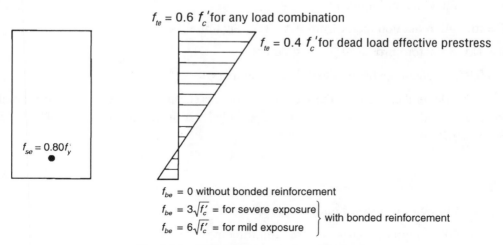

$f_{te} = 0.6\ f'_c$ for any load combination

$f_{te} = 0.4\ f'_c$ for dead load effective prestress

$f_{se} = 0.80f'_y$

$f_{be} = 0$ without bonded reinforcement

$f_{be} = 3\sqrt{f'_c}$ = for severe exposure

$f_{be} = 6\sqrt{f'_c}$ = for mild exposure

} with bonded reinforcement

Fig. 3-24.   Allowable stresses at working load

Ultimate Limit State in Flexure

The nominal strength of a prestressed rectangular flexural member is derived in a manner similar to that used for a member that is not prestressed. It is given by AASHTO Equation (9-13) as

$$M_n = A_s^* f_{su}^* d (1 - 0.60 \, p^* f_{su}^* / f_c')$$

where

A_s^* = area of prestressing steel

d = depth to centroid of prestressing force

$p^* = A_s^*/bd$ = ratio of prestressing steel

f_{su}^* = average stress in prestressing steel at ultimate load

Provided the effective prestress in the tendons after losses f_{se}, is not less than half the tensile strength f_s', the stress in bonded tendons at ultimate load is given by AASHTO Equation (9-17) as

$$(d_t \, / \, d)(pf_{sy} \, / \, f_c') + p^* f_{su}^* \, / \, f_c' - p \, 'f_y' \, / \, f_c' \le 0.36 \beta_1$$

where, β_1 = compression zone factor given in AASHTO Section 8.16.2.7

γ^* = prestressing steel factor given in AASHTO Section 9.1.2

Similarly, for unbonded tendons, the stress in the tendons at ultimate load is given by AASHTO Equation (9-18) as

$$f_{su}^* = f_{se} + 900(d - y_u)l_e$$

where

$l_e = l_i/(1 + 0.5N_s)$ = effective tendon length

y_u = depth to neutral axis at tendon yielding

l_i = tendon length

N_s = number of support hinges crossed by the tendon

The design flexural capacity is given by

$$\phi M_n \ge M_u$$

where

M_u = applied factored moment

ϕ = strength reduction factor from AASHTO Section 9.14

  = 1.0 for factory produced precast members

  = 0.95 for post-tensioned cast-in-place members.

The flexural capacity of the section may be increased by means of non-prestressed reinforcement, and the additional tensile force provided at ultimate load is specified by AASHTO Section 9.19 as

$$T_u = A_s f_y'$$

where

A_s = area of non-prestressed reinforcement

f_y' = yield strength of non-prestressed reinforcement.

To ensure adequate warning of impending failure, with considerable yielding of the steel before compressive failure of the concrete, the reinforcement index for rectangular sections is given by AASHTO Equation (9-24) as

$$(d_t/d)(pf_{sy}/f_c') + p^*f_{su}^*/f_c' - p'f_y'/f_c' \le 0.36\, \beta_1$$

where

$p = A_s/bd_t$ = ratio of non-prestressed tension reinforcement

d_t = depth to centroid of non-prestressed reinforcement

$p^* = A_s^*/bd$ = ratio of prestressing steel

$p' = A_s'/bd$ = ratio of compression steel

f_y' = yield stress in compression of compression reinforcement

Bridge
Structures

The design flexural strength of a prestressed rectangular section with non-prestressed reinforcement is given by AASHTO Equation (9-13a) and the stress in the prestressing steel is given by AASHTO Equation (9-17a).

To prevent sudden, premature tensile failure of the steel, AASHTO Section 9.18.2 specifies that

$$\phi M_n \ge 1.2\, M_{cr}^*$$

where, M_{cr}^*, the cracking moment strength, is computed by linear elastic theory using a modulus of rupture given by AASHTO Section 9.15.2 as

$$f_r = 7.5\sqrt{f_c'}$$

Ultimate Limit State in Shear

At the ends of a beam, the critical section for shear is defined in AASHTO Section 9.20.1.4 as being located a distance $h/2$ from the support.

The factored applied shear force acting on a member, in accordance with AASHTO Equation (9-26), is resisted by the combined shear strength of the concrete and the shear reinforcement and is given by

$$V_u = \phi V_c + \phi V_s$$

where

$\phi = 0.90$ = strength reduction factor from AASHTO Section 9.14

V_c = nominal shear capacity provided by the concrete

V_s = nominal shear capacity provided by the web reinforcement.

Shear failure in prestressed concrete beams may occur in two different modes, as shown in Fig. 3-25. In a zone which is cracked in flexure, a flexural crack may develop into an inclined crack and eventually cause a shear failure. The position of this crack varies but is usually located at a distance of half the effective depth from the point of maximum moment. In a zone where the bending moment is less than the cracking moment, flexural cracks do not occur and shear failure is caused when the principal tensile stress exceeds the tensile strength of the concrete, thus producing a web shear crack. Since web shear cracks may also occur in a zone cracked in flexure, it is necessary to design for web shear cracking throughout the beam.

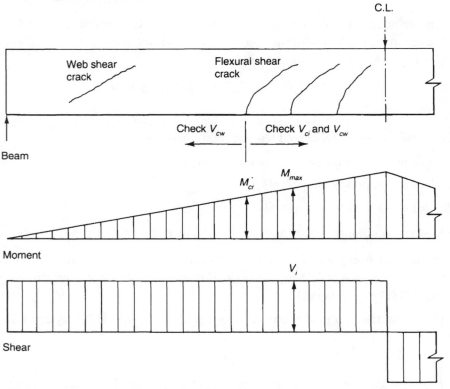

Fig. 3-25.   Shear failure modes

The nominal web shear capacity of the concrete is given by AASHTO Equation (9-29) as

$$V_{cw} = b'd\left(3.5\sqrt{f_c'} + 0.3f_{pc}\right) + V_p$$

where

b' = web width

d = depth to centroid of prestressing force $\geq 0.80h$

f_{pc} = concrete compressive stress at centroid of section due to effective prestress force

V_p = vertical component of effective prestress force

In a pretensioned member, the prestress force varies linearly from zero at the end of the member to a maximum at the end of the transfer length which is 50 diameters for strand and 100 diameters for wire. Hence, if the critical section at $h/2$ from the support lies within the transfer length, the value of f_{pc} is reduced accordingly.

When the centroid of a flange section lies within the flange, the value of f_{pc} is determined at the junction of the web and the flange and includes the compressive stresses due to applied dead and live loads.

The nominal flexural shear capacity of the concrete is given by AASHTO Equation (9-27) as

$$V_{ci} = 0.60\,b'd\sqrt{f_c'} + V_d + V_i M_{cr}' / M_{max} \geq 1.70b'd\sqrt{f_c'}$$

where

V_d = shear force due to unfactored dead load.

M_{max} = factored moment due to externally applied loads

V_i = factored shear, due to externally applied loads, occurring simultaneously with M_{max}

M'_{cr} = the unfactored moment, due to externally applied loads, required to produce cracking

$$= (\phi f_r + f_{pe} - f_d) I / \bar{y}$$

$$= (0.8 \times 7.5 \sqrt{f'_c} + f_{pe} - f_d) I / \bar{y}$$

$$= (6.0 \sqrt{f'_c} + f_{pe} - f_d) I / \bar{y}.$$

$\phi = 0.8$ = reduction factor for shrinkage and variable concrete quality

$f_r = 7.5 \sqrt{f'_c}$ = modulus of rupture

f_{pe} = concrete compressive stress at extreme fibre due to effective prestress force.

f_d = stress at extreme fibre due to unfactored dead load

I = moment of inertia of section

$\bar{y}$ = distance of section centroid from extreme fibre

The nominal shear capacity provided by the concrete V_c, is taken as the lesser of the values of V_{cw} and V_{ci}. The provision of shear reinforcement is governed by the requirements of AASHTO Section 9.20.1 and Section 9.20.3.

No shear reinforcement is required when the applied factored shear force is given by

$$V_u < V_c / 2$$

Nominal shear reinforcement is required when

$$\phi V_c / 2 \leq V_u \leq \phi V_c$$

and the nominal area of shear reinforcement is given by AASHTO Equation (9-31) as

$$A_{v(min)} = 50 \, b' \, s / f_{sy}$$

where

f_{sy} = yield strength of shear reinforcement $\leq 60,000$ pounds per square inch

Designed shear reinforcement is required when

$$\phi V_c < V_u \leq \phi(V_c + 8 b' d \sqrt{f'_c})$$

and the area of shear reinforcement required is given by AASHTO Equation (9-30) as

$$A_v = s V_s / d f_{sy}$$

where

s = spacing of shear reinforcement

$V_s = V_u / \phi - V_c$ = nominal capacity of shear reinforcement

When the required nominal capacity is

$$V_s \leq 4 b' d \sqrt{f'_c}$$

the reinforcement spacing is given by

$0.75d \geq s \leq 24$ inches.

When the required nominal capacity is

$$4 b' d \sqrt{f'_c} < V_s \leq 8 b' d \sqrt{f'_c}$$

the reinforcement spacing is given by

$0.375\, d \geq s \leq 12$ inches.

The dimensions of the section or the concrete strength must be increased to ensure that

$$V_s \leq 8\, b'd\sqrt{f_c'}$$

Example 5

The prestressed concrete cast-in-place Tee-beam superstructure, shown in Fig. 3-26, has a concrete strength of 5,000 pounds per square inch. The cable centroid, as shown, is parabolic in shape and the final prestressing force after all losses is 500 kip. At section x-x, which is 26 inches from the support, the unfactored dead load shear and moment are 21 kip and 45 kip feet and the unfactored shear and moment due to live load plus impact are 56 kip and 121 kip feet. Determine the required spacing of Number 3, Grade 60 stirrups.

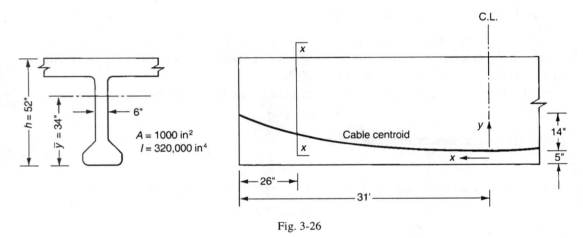

Fig. 3-26

Solution

The equation of the parabolic cable profile is

$y = rx^2/a^2$

where, $r = 14$ inches = cable drape

a = half length of cable = 31 feet.

The rise of the cable at section x-x is given by

$y = 14(31 - 26/2)^2/31^2$

  $= 12$ inches.

The cable eccentricity at section x-x is

$e = \bar{y} - 5 - y$

  $= 34 - 5 - 12$

  $= 17$ inches

The effective depth at section x-x is given by

$d = h - 5 - y$

  $= 52 - 5 - 12$

= 0.80h minimum

= 41.6 inches

The equation of the slope of the parabolic cable is

$dy/dx = 2rx/a^2$

The slope of the cable at section x-x is given by

$dy/dx = 2 \times 14(31 - 26/12)/(12 \times 31^2)$

$= 0.070$

The vertical component of the final prestressing force at section x-x is

$V_p = P_e \, dy/dx$

$= 500 \times 0.070$

$= 35$ kips

The compressive stress at the centroid of the cross-section, at section x-x, due to the final prestressing force is

$f_{pc} = P_e/A$

$= 500/1000$

$= 0.500$ kips per square inch

    The nominal web shear capacity of the section is given by AASHTO Equation (9-29) as

$V_{cw} = b'd \, (3.5 \sqrt{f_c'} + 0.3f_{pc}) + V_p$

$= 6 \times 41.6 \, (3.5 \sqrt{5000} + 0.3 \times 500)/1000 + 35$

$= 134$ kips

The compressive stress at the bottom face of the cross-section, at section x-x, due to the final prestressing force, is

$f_{pe} = P_e/A + eP_e \bar{y}/I$

$= 500/1000 + 17 \times 500 \times 34/320,000$

$= 1.403$ kips per square inch.

    The tensile stress at the bottom face of the cross-section, at section x-x, due to the unfactored dead load is

$f_d = M_d \bar{y}/I$

$= 45 \times 12 \times 34/320,000$

$= 0.057$ kips per square inch.

    The unfactored moment, due to externally applied loads, required to produce cracking at section x-x is given by AASHTO Equation (9-28) as

$M_{cr}' = (6.0 \sqrt{f_c'} + f_{pe} - f_d)I/\bar{y}$

$= (6.0 \sqrt{5000} + 1403 - 57) \, 320,000/(12,000 \times 34)$

$= 1388$ kip feet

The unfactored live load moment at section x-x is

$M_L = 121$ kip feet

$< M'_{cr}$

Hence, cracking does not occur at section x-x and it is unnecessary to determine the nominal flexural shear capacity. The nominal shear capacity provided by the concrete is, then,

$V_c = V_{cw}$

$= 134$ kips

The factored shear force acting on section x-x is given by AASHTO Equation (3-10) as

$V_u = \gamma[\beta_D D + \beta_L(L + I)]$

$= 1.3\ (1 \times 21 + 1.67 \times 56)$

$= 149$ kips

The nominal shear strength required from shear reinforcement is given by AASHTO Equation (9-26) as

$V_s = V_u/\phi - V_c$

$= 149/0.9 - 134$

$= 31.6$ kips

$< 4\,b'd\sqrt{f'_c}$

Hence, the stirrup spacing is given by

$s \le 0.75d$

$\le 0.75 \times 41.6$

≤ 24 inches . . . governs

The required spacing of Number 3 stirrups is given by AASHTO Equation (9-30) as

$s = A_v df_{sy}/V_s$

$= 0.22 \times 41.6 \times 60/16.6$

$= 24$ inches maximum governs

To satisfy the minimum area of shear reinforcement given by AASHTO Equation (9-31) the required spacing of Number 3 stirrups is

$s = A_v f_{sy}/50\,b'$

$= 0.22 \times 60,000/(50 \times 6)$

$= 24$ inches maximum governs.

Loss of Prestress

The determination of friction losses in AASHTO Section 9.16.1 is identical with the method presented in the ACI Code[23] for building structures, and this is covered in Chapter 2 of this text. Values for the wobble and friction coefficients are given in AASHTO Section 9.16.1 and alternative values are proposed in the ACI Code[24] for bridge structures. The effect of friction losses on cable stress, in a post-tensioned member, is illustrated in Fig. 2-17 of this text.

In a post-tensioned member, allowance is necessary for anchor set, or draw-in, which occurs in the anchorage system when the prestressing force is transferred from the tensioning equipment to the anchorage. The magnitude of this loss depends on the prestressing system employed and is particularly pronounced with short members. The effect of wedge slip on cable stress, in a post-tensioned member, is illustrated in Fig. 2-18 of this text. Initial over-stressing of the tendon to $0.90f_y^*$, as shown in Fig. 3-23, increases the stress in the interior of the member. However, the loss at the stressing, or live, end still occurs.

The remaining loss of prestress, due to all other causes, is summarized in AASHTO Equation (9-3) as

$$\Delta f_s = SH + ES + CR_c + CR_s$$

where

SH = loss due to concrete shrinkage

ES = loss due to elastic shortening

CR_c = loss due to creep of concrete

CR_s = loss due to relaxation of prestressing steel

An approximate estimate of this loss for normal exposure conditions, span length not exceeding 120 feet, and normal weight concrete may be obtained from AASHTO Table 9.16.2.2. For 5000 pounds per square inch concrete, the value given for pretensioned strand is 45,000 pounds per square inch and for post-tensioned strand is 33,000 pounds per square inch. More accurate estimates may be obtained as detailed in AASHTO Section 9.16.2.

Shrinkage loss is determined by AASHTO Equation (9-4) and (9-5) as

$$SH = 17,000 - 150RH \dots \text{pretensioned member}$$

$$SH = 0.80(17,000 - 150RH) \dots \text{post-tensioned member}$$

where

RH = percentage mean annual ambient relative humidity

Elastic loss is determined from AASHTO Equations (9-6) and (9-7) as

$$ES = n_i f_{cir} \dots \text{pretensioned member}$$

$$ES = n_i f_{cir}/2 \dots \text{post-tensioned member}$$

where

$n_i = E_s/E_{ci}$ = modular ratio at transfer

$E_s = 28 \times 10^6$ pounds per square inch = modulus of elasticity of prestressing steel

$E_{ci} = 33\, w^{3/2} \sqrt{f'_{ci}}$ = modulus of elasticity of concrete at transfer

w = density of concrete

f'_{ci} = concrete strength at transfer

f_{cir} = concrete stress at level of tendon centroid immediately after transfer.

Creep loss is determined from AASHTO Equation (9-9) as

$$CR_c = 12f_{cir} - 7f_{cds}$$

where

f_{cds} = concrete stress at level of tendon centroid due to all sustained dead load except the dead load present at transfer.

Relaxation loss for stress relieved strand is determined from AASHTO Equation (9-10) and (9-11) as

$CR_s = 20,000 - 0.4ES - 0.2(SH + CR_c)$ . . . pretensioned member

$CR_s = 20,000 - 0.4ES - 0.2(SH + CR_c) - 0.3FR$ . . . post-tensioned member

where

FR = friction stress reduction below $0.70 f_s'$

Example 6

Determine the prestressing losses in the post-tensioned member shown in Fig. 3-27. The concrete strength at 28 days is 5000 pounds per square inch, and at transfer is 3500 pounds per square inch. The total area of the prestressing tendons is 2.5 square inches, and the stress-relieved strand has a specified tensile strength of 270 kip per square inch, a nominal yield stress of 230 kip per square inch and an anchor set of 0.125 inches. Stressing is effected from one end of the member only, and the applicable friction coefficient is 0.25, and the wobble coefficient is 0.0015 per foot. The bridge is constructed in an area with an annual average ambient relative humidity of 65%.

Additional superimposed dead load due to diaphragms, surfacing, curbs and parapets may be neglected.

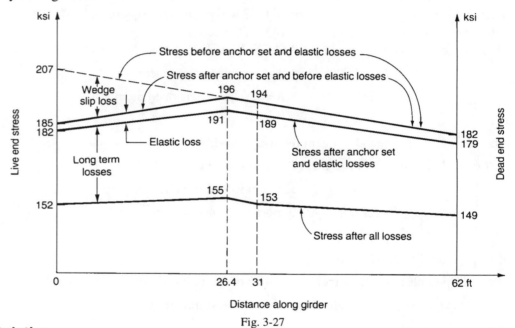

Distance along girder

Fig. 3-27

Solution

The nominal radius of the cable profile is

$R = a^2/2r$

$= 12 \times 31^2/(2 \times 14)$

$= 412$ feet.

The cable length, along the curve, from the end of the member to center span is

$L_c = 2rR/a$

$= 2 \times 14 \times 412/(12 \times 31)$

$= 31.01$ feet.

The friction loss may be determined over the two sections, from the live end to center span and from center span to the dead end.

The maximum allowable jacking stress is given by AASHTO Section 9.15.1 as

$$f_{s0} = 0.90 \, f_y^*$$

$$= 0.90 \times 230$$

$$= 207 \text{ kips per square inch.}$$

The combined loss factor due to friction and wobble is

$$c = K + \mu/R$$

$$= 0.0015 + 0.25/412$$

$$= 21 \times 10^{-4} \text{ per foot}$$

The stress at a distance x along the cable is

$$f_{sx} = f_{s0} \, e^{-cx}$$

At center span,

$$x = 31.01 \text{ feet}$$

hence, $f_{s(31.01)} = 207 \times 0.937$

$$= 194 \text{ kips per square inch}$$

At the dead end,

$$x = 62.02 \text{ feet}$$

hence, $f_{s(62.02)} = 207 \times 0.878$

$$= 182 \text{ kips per square inch.}$$

These stress values are plotted on Fig. 3-27.

The magnitude of the anchor set is given as

$$\Delta = 0.125 \text{ inches}$$

AASHTO Section 9.16.2 specifies that the modulus of elasticity of prestressing strand is

$$E_s = 28 \times 10^3 \text{ kips per square inch}$$

The stress loss per foot is given by

$$m = (207 - 194)/31$$

$$= 0.419 \text{ kips per square inch per foot}$$

The length of cable at the live end, affected by the wedge pull-in, is derived in Chapter 2 of this text as

$$x_s = \sqrt{\Delta E_s / 12m}$$

$$= \sqrt{0.125 \times 28 \times 10^3 / (12 \times 0.419)}$$

$$= 26.4 \text{ feet}$$

The stress at the end of the pull-in zone is

$$f_{s(26.4)} = f_{s0} - mx_s$$

$$= 207 - 0.419 \times 26.4$$

$$= 196 \text{ kips per square inch}$$

Bridge Structures

The stress at the live anchorage is

$$f_{s(\text{anc})} = f_{s0} - 2mx_s$$

$$= 185 \text{ kips per square inch.}$$

The above stresses, which are fictitious stresses prior to considering elastic losses, are plotted on Fig. 3-27.

AASHTO Section 9.16.2 specifies that the modulus of elasticity of the concrete at transfer is

$$E_{ci} = 33 \; w^{3/2} \; \sqrt{f_{ci}'}$$

$$= 33 \times 150^{3/2} \; \sqrt{3500} / 1000$$

$$= 3590 \text{ kips per square inch}$$

Hence the modular ratio at transfer is

$$n_i = E_s / E_{ci}$$

$$= 28{,}000/3590$$

$$= 7.8$$

The total self weight of the girder is

$$W_G = 0.150 LA / 144$$

$$= 0.150 \times 62 \times 1000 / 144$$

$$= 64.58 \text{ kips}$$

The bending moment at mid span due to the self weight of the girder is

$$M_G = W_G L / 8$$

$$= 64.58 \times 62 \times 12 / 8$$

$$= 6006 \text{ kip inches}$$

The concrete stress at mid span, at the level of the tendon centroid, due to the girder self weight is

$$f_{cG} = -e M_G / I$$

$$= -29 \times 6006 / 320{,}000$$

$$= -0.544 \text{ kips per square inch}$$

Assuming a 5 kip per square inch loss of prestress in the tendon due to elastic shortening of the concrete section, the initial prestressing force at mid span is

$$P_i = A_s^* f_{si}$$

$$= A_s^* (f_{s(31.01)} - 5)$$

$$= 2.5 \, (194 - 5)$$

$$= 473 \text{ kip.}$$

The concrete stress at mid span, at the level of the tendon centroid, due to the initial prestressing force is

$$f_{cip} = P_i (1/A + e^2/I)$$

$$= 473(1/1000 + 29^2/320{,}000)$$

$$= 1.716 \text{ kips per square inch.}$$

The concrete stress at mid span, at the level of the tendon centroid, immediately after transfer, and allowing for loss of prestress in the tendon due to elastic shortening of the concrete section is

$$f_{cir} = f_{cip} + f_{cG}$$
$$= 1.716 - 0.544$$
$$= 1.172 \text{ kips per square inch.}$$

The elastic shortening caused by this stress level produces a loss in prestress in the tendon of

$$ES = n_i f_{cir}/2$$
$$= 7.8 \times 1.172/2$$
$$= 4.6 \text{ kips per square inch}$$
$$\approx 5 \text{ kips per square inch, assumed . . . satisfactory.}$$

Assuming a 3 kip per square inch loss of prestress in the tendon, due to elastic shortening, at the live anchorage, the initial prestressing force at the live anchor is

$$P_i = A_s^* f_{si}$$
$$= A_s^* (f_{s(anc)} - 3)$$
$$= 2.5(185 - 3)$$
$$= 455 \text{ kips}$$

The concrete stress at the live anchorage, at the level of the tendon centroid, due to this initial prestressing force is

$$f_{cip} = f_{cir} = P_i(1/A + e^2/I)$$
$$= 455(1/1000 + 15^2/320,000)$$
$$= 0.775 \text{ kips per square inch}$$

The elastic shortening caused by this stress level produces a loss in prestress in the tendon of

$$ES = n_i f_{cir}/2$$
$$= 7.8 \times 0.775/2$$
$$= 3 \text{ kips per square inch}$$
$$= \text{value initially assumed . . . satisfactory.}$$

Hence, immediately after elastic losses have occurred, the stresses in the prestressing tendon are:

at the live anchorage, $f_s = 185 - 3 = 182$ kips per square inch

at mid span, $f_s = 194 - 5 = 189$ kips per square inch

at the dead anchorage, $f_s = 182 - 3 = 179$ kips per square inch

at the end of the pull–in zone, $f_s = 196 - 5 = 191$ kips per square inch.

These stress values are plotted in Fig. 3-27.

The allowable stress at the live anchorage, immediately after transfer, is given by AASHTO Section 9.15.1 as

$$f_{s(all)} = 0.70 \, f_s' \ldots \text{stress relieved strand}$$
$$= 0.70 \times 270$$

Bridge Structures

= 189 kips per square inch

> 182 . . . satisfactory.

The allowable stress at the end of the pull-in zone, immediately after transfer, is given by AASHTO Section 9.15.1 as

$$f_{s(all)} = 0.83f_y^*$$

$$= 0.83 \times 230$$

$$= 191 \text{ kips per square inch}$$

$$= 191, \text{ actual value . . . satisfactory.}$$

The shrinkage loss is determined from AASHTO Equation (9-5) as

$$SH = 0.80(17,000 - 150RH)/1000$$

$$= 0.80(17,0000 - 150 \times 65)/1000$$

$$= 6 \text{ kips per square inch.}$$

The creep loss is determined from AASHTO Equation (9-9) as

$$CR_c = 12f_{cir} - 7f_{cds} = 12f_{cir}$$

since additional superimposed dead load may be neglected.

At mid span at transfer, the concrete stress at the level of the tendon, allowing for elastic shortening and self weight effects, is

$$f_{cir} = 1.172 \text{ kips per square inch}$$

Hence, the creep loss at mid span is

$$CR_c = 12f_{cir}$$

$$= 12 \times 1.172$$

$$= 14 \text{ kips per square inch.}$$

Similarly, at the live anchorage, the creep loss is given by

$$CR_c = 12f_{cir}$$

$$= 12 \times 0.775$$

$$= 9 \text{ kips per square inch}$$

At mid span, the friction stress reduction below $0.70\,f_s'$ is

$$FR = 0.70\,f_s' - f_{s(31.01)}$$

$$= 189 - 194$$

$$= -5 \text{ kips per square inch}$$

At the live anchorage, the friction stress reduction below $0.70\,f_s'$ is

$$FR = 0.70\,f_s' - f_{s(anc)}$$

$$= 189 - 185$$

$$= 4 \text{ kips per square inch}$$

The creep loss, at mid span, is determined from AASHTO Equation (9-11) as

$$CR_s = 20 - 0.4ES - 0.2(SH + CR_c) - 0.3FR$$

$$= 20 - 0.4 \times 5 - 0.2(6 + 14) - 0.3(-5)$$

$$= 16 \text{ kips per square inch.}$$

The creep loss, at the live anchorage, is given by

$$CR_s = 20 - 0.4 \times 3 - 0.2(6 + 9) - 0.3 \times 4$$

$$= 15 \text{ kips per square inch.}$$

Hence, the total long term loss at mid span is

$$\Delta f_e = SH + CR_c + CR_s$$

$$= 6 + 14 + 16$$

$$= 36 \text{ kips per square inch.}$$

At the live anchorage, the long term loss is

$$\Delta f_e = 6 + 9 + 15$$

$$= 30 \text{ kips per square inch.}$$

These values are plotted on Fig. 3-27.

The maximum allowable stress at service load is given by AASHTO Section 9.15.1 as

$$f_{s(all)} = 0.80 f_y^*$$

$$= 0.80 \times 230$$

$$= 184$$

$$> 155 \dots \text{ satisfactory.}$$

Composite Construction

Composite construction consists of a member in which cast-in-place concrete is added to a precast concrete unit. Provided that measures are employed to prevent excessive slip across the interface between the two concretes, complete interaction may be assumed, and the composite member designed as a monolithic member. Typically, the precast concrete unit is prestressed and the cast-in-site concrete is reinforced. In accordance with AASHTO Section 9.20.4, specific values of the horizontal shear resistance at the interface may be assumed if the contact surfaces are roughened and vertical ties are provided over the span length.

Two methods of construction are employed, propped and unpropped. In the unpropped method, the precast unit acts as formwork for, or supports the formwork for, the cast-in-place concrete. For this technique, the precast unit is designed to support its self weight, formwork if required, and the weight of the cast-in-situ concrete. In the propped method, the precast unit is supported during the placing and curing of the cast-in-place concrete.

Requirements at the transfer limit state for the precast section are identical with those for an integral member. Elastic losses occur at this time, together with friction losses if the precast girder is post-tensioned, giving an initial prestressing force of P_i. The self weight moment of the girder, M_G, also acts on the precasting unit at this time, as shown in Fig. 3-28 and 3-29, and the initial stresses are indicated.

Conditions at the serviceability state for unpropped construction are shown in Fig. 3-28, and for propped construction are shown in Fig. 3-29.

For unpropped construction, the precast girder supports the weight of the shuttering, which produces a sagging bending moment M_S, and also the weight of the flange concrete,

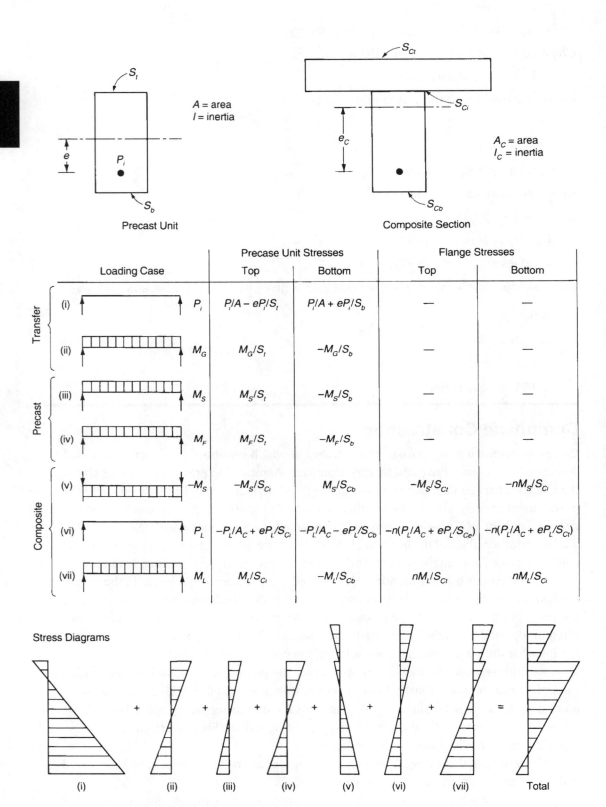

Fig. 3-28. Unpropped composite construction

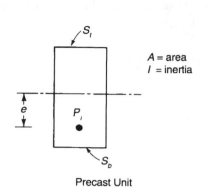

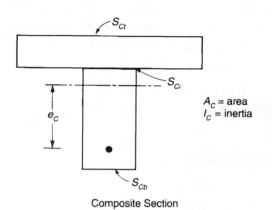

A = area
I = inertia

A_c = area
I_c = inertia

**Bridge
Structures**

Precast Unit   Composite Section

	Loading Case		Precast Unit Stresses		Flange Stresses	
			Top	Bottom	Top	Bottom
Transfer	(i)	P_i	$P_i/A - eP_i/S_t$	$P_i/A + eP_i/S_b$	—	—
	(ii)	M_G	M_G/S_t	$-M_G/S_b$	—	—
Precast	(iii) Prop		0	0	—	—
	(iv) Flange and Shutter		0	0	—	—
Composite	(v) −Shuttering		0	0	0	0
	(vi)	M_F	M_F/S_{Ci}	$-M_F/S_{Cb}$	nM_F/S_{Ct}	nM_F/S_{Ci}
	(vii)	P_L	$-P_L/A_c + eP_L/S_{Ci}$	$-P_L/A_c - eP_L/S_{Cb}$	$-n(P_L/A_c + eP_L/S_{Ct})$	$-n(P_L/A_c + eP_L/S_{Ci})$
	(viii)	M_L	M_L/S_{Ci}	$-M_L/S_{Cb}$	nM_L/S_{Ci}	nM_L/S_{Ci}

Stress Diagrams

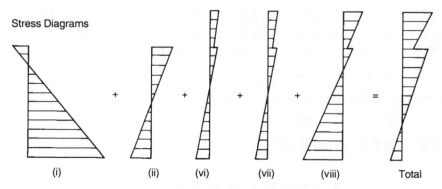

Fig. 3-29.   Propped composite construction

which produces a bending moment M_F, and the resultant stresses are indicated in Fig. 3-28. After curing, the flange and precast girder form a composite section, and the effective width of the flange is given by AASHTO Section 8.10.1 as the lesser of

    (i) $L/4$

    (ii) $b_w + 6h_f$

    (iii) S

where

L = span length

b_w = web thickness

h_f = flange depth

S = rib spacing.

In addition, when the 28-day compressive strengths of the precast section and the flange differ, the effective flange width is transformed by dividing by the modular ratio

$$n = E_f/E_r$$

where

E_f = modulus of elasticity of the flange concrete

E_r = modulus of elasticity of the girder concrete.

    Removal of the shuttering produces a hogging moment $-M_s$, which acts on the composite section giving the stresses indicated.

    Long term prestressing losses, which cause a loss of prestressing force P_L, now act on the composite section producing the stresses indicated.

    Additional superimposed dead load, live load, and impact, which produce the bending moment M_L, act on the composite section, producing the stresses indicated.

    In propped construction, the soffit of the precast unit is supported on falsework prior to placing the shuttering and casting the flange. No additional stresses are produced in the precast unit since the weight of the shuttering and the flange are carried by the falsework. Similarly, removal of the shuttering produces no stresses. On removing the falsework, the weight of the flange concrete, which produces a bending moment M_F, acts on the composite section and causes the stresses indicated in Fig. 3-29. Long term prestressing losses, additional superimposed dead load, live load and impact, also act on the composite section producing the stresses indicated.

    At the serviceability limit state, the effect of differential shrinkage should also be considered. When the flange is cast on the precast unit, much of the shrinkage has already occurred in the unit and the additional shrinkage of the flange is restrained by the precast unit. As shown in Fig. 3-30, this produces a force at the center of the flange of

$$F_f = \varepsilon A_f E_f$$

and a moment about the centroid of the composite section of

$$M_f = e_f F_f.$$

    The resultant stresses in the composite section are reduced[26] due to creep by the creep factor

$$r = [1 - \exp(-E_f c)]/E_f c$$

where, ε = differential shrinkage strain

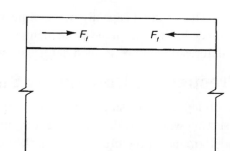

Fig. 3-30.   Effects of differential shrinkage

A_f = actual area of flange

E_f = modulus of elasticity of flange concrete

c = unit creep strain

Due to these forces, the stress at the top of the composite section is

$$f_{Ct} = r[n(F_f/A_C + M_f/S_{Ct}) - \varepsilon E_f]$$

and at the bottom of the composite section

$$f_{Cb} = r(F_f/A_C - M_f/S_{Cb})$$

where, compressive stress is positive.

A multi-span bridge composed of simply supported precast units may be made continuous by providing a cast-in-place concrete diaphragm at each support, as shown in Fig. 3-31(i). The superstructure is now continuous for live loads and additional superimposed dead load, and continuity reinforcement is provided in the slab to resist the negative moment at the support. In addition, AASHTO Section 9.7.2 requires the provision of reinforcement at the bottom of the connection, in order to provide restraint for positive moment developed there.

As shown in Fig. 3-31(ii), an unrestrained composite beam tends to hog with time due to differential shrinkage and creep caused by prestressing force and self weight. When restraint is imposed on the beam ends, a positive moment is developed as shown. The effects of this may be determined by moment distribution techniques.[27]

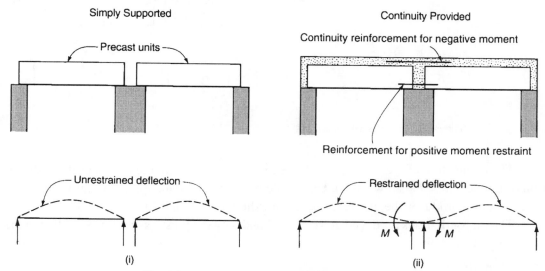

Fig. 3-31.   Creep and shrinkage effects at connections

The nominal capacity of a composite section, in flexure and shear, is determined at the ultimate limit state, in the same manner as an integral member. Differential shrinkage effects, which result from restrained deformations, do not affect ultimate limit state conditions.

Prestressed Continuous Structures

Applying a prestressing force P to a statically determinate structure produces a moment Pe at a section where e is the cable eccentricity at that section. Applying a prestressing force to an indeterminate structure tends to deflect the structure. This produces indeterminate reactions at the redundant supports which cause secondary moments in the structure. By removing the redundant supports, to produce the cut-back structure, and applying the prestressing force P, the primary moments Pe are again produced. The total moment at any section may then be obtained by the superposition of the primary and secondary moments.

The application of the prestressing force P, with an eccentricity e, to the continuous beam shown in Fig. 3-32 results, in general, in the production of indeterminate reactions at the supports. The prestressing force tends to deflect the beam, which is restrained against lateral displacement by the supports, and this causes secondary moments. The resultant line of thrust no longer coincides with the cable profile, as is the case with a statically determinate beam. The indeterminate moments M_2 and M_3 at the supports are taken as the redundants and releases introduced at 2 and 3 to produce the cut-back structure.

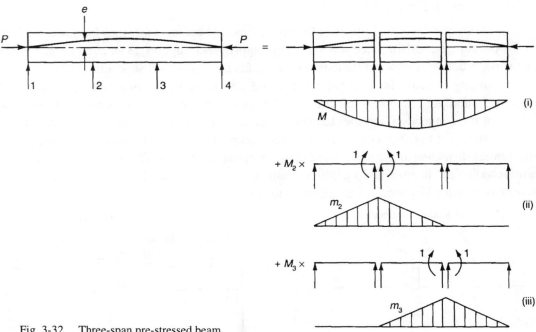

Fig. 3-32.   Three-span pre-stressed beam

The application of the prestressing force to the cut-back structure produces the distribution of moment $M = Pe$ shown at (i). Unit value of each redundant applied in turn to the cut-back structure produces the moments m_2 and m_3 shown at (ii) and (iii).

The discontinuities produced at Releases 2 and 3 by the prestressing force applied to the cut-back structure are

$$\theta_2 = \int Pem_2 \, ds/EI$$

$\theta_3 = \int Pem_3 \, ds/EI.$

The discontinuities produced at Release 2 by unit values of M_2 and M_3 applied, in turn, to the cut-back structure are

$f_{22} = \int m_2^2 \, ds/EI$

$f_{23} = \int m_2 m_3 \, ds/EI.$

The discontinuities produced at Release 3 by unit values of M_2 and M_3 applied, in turn, to the cut-back structure are

$f_{32} = \int m_2 m_3 \, ds/EI$

$f_{33} = \int m_3^2 \, ds/EI.$

Since there are no discontinuities in the original structure at the positions of the releases

$$\begin{bmatrix} \int m_2^2 ds/EI & \int m_2 m_3 ds/EI \\ \int m_2 m_3 ds/EI & \int m_3^2 ds/EI \end{bmatrix} \begin{bmatrix} M_2 \\ M_3 \end{bmatrix} = - \begin{bmatrix} \int Pem_2 \, ds/EI \\ \int Pem_3 \, ds/EI \end{bmatrix}$$

Expanding this expression gives

$\int (Pe + M_2 m_2 + M_3 m_3) m_2 \, ds/EI = 0$

$\int (Pe + M_2 m_2 + M_3 m_3) m_3 \, ds/EI = 0$

The final distribution of moment in the beam due to the prestressing force and secondary effects is

$Pe' = Pe + M_2 m_2 + M_3 m_3$

where the effective cable eccentricity is

$e' = e + M_2 m_2/P + M_3 m_3/P$

and the line of thrust has been displaced by an amount

$M_2 m_2/P + M_3 m_3/P.$

Pe = primary moment

$M_2 m_2 + M_3 m_3$ = secondary moment

A cable with an initial eccentricity e' produces discontinuities at Releases 2 and 3 of

$\theta_2 = \int Pe'm_2 \, ds/EI$

$\quad = \int (Pe + M_2 m_2 + M_3 m_3) m_2 \, ds/EI$

$\quad = 0.$

$\theta_3 = \int Pe'm_3 \, ds/EI$

$\quad = \int (Pe + M_2 m_2 + M_3 m_3) m_3 \, ds/EI$

$\quad = 0.$

Thus, no secondary moments are produced on tensioning this cable and this is the concordant cable. The resultant line of thrust coincides with the cable profile.

Stressing the cable in the two-hinged portal frame shown in Fig. 3-33 produces a horizontal thrust H at the supports. The cut-back structure is obtained by introducing a release at 1. The application of the prestressing force to the cut-back structure produces the distribution of moment $M = Pe$ shown at (i). Unit value of H applied to the cut-back structure produces the moment m shown at (ii). Equating the discontinuities at the releases to zero gives

$H \int m^2 \, ds/EI = - \int Pem \, ds/EI$

and H may be determined.

The final distribution of moment in the frame due to the prestressing force and secondary effects is

$Pe' = Pe + Hm$

where the effective cable eccentricity is

$e' = e + Hm/P$

This is the expression for the concordant cable profile which is shown at (iii).

Pe = primary moment

Hm = secondary moment.

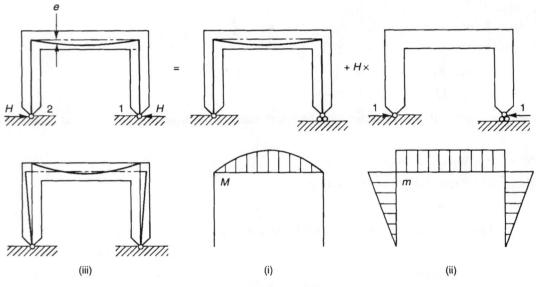

Fig. 3-33.   Prestressed frame

Example 7

The symmetrical, two-span beam shown in Fig. 3-34 is prestressed with a cable placed to a parabolic profile with an eccentricity e_0 at the central support. Determine the position of the resultant line of thrust.

Solution

The reaction at the central support V is taken as the redundant and a release introduced at 2 to produce the cut-back structure.

The application of the prestressing force to the cut-back structure produces the distribution of moment $M = Pe$ shown at (i). Unit value of V applied to the cut-back structure produces the moment m shown at (ii). Equating the discontinuities at the release to zero gives

$V \int m^2 \, ds/EI = - \int Pem \, ds/EI$

$Vl^3/6 = - 5l^2 Pe_0/12.$

The effective cable eccentricity is

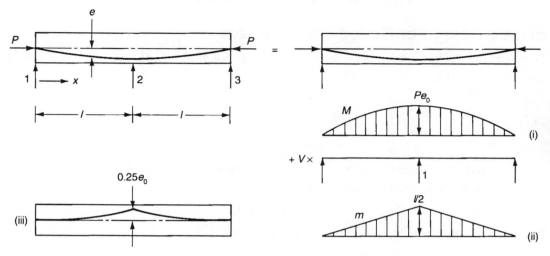

Fig. 3-34

$$e' = e - 5xe_0/4l$$

and the resultant line of thrust is shown at (iii).

When the value of EI varies along the length of a member, the integrals involved in the application of the compatibility or flexibility matrix method may be evaluated by Simpson's rule. The member is divided into an even number of segments of equal length, and the integral of the function shown in Fig. 3-35 is given by

$$\int g \, dx = s(g_1 + 4g_2 + 2g_3 + 4g_4 + \ldots + 2g_{n-2} + 4g_{n-1} + g_n)/3$$

An alternative design procedure[28,29] is available which utilizes the load balancing technique. The equivalent lateral loads, produced by the cable profile, are obtained from Table 2-21 of this book and the resultant moments determined by moment distribution.

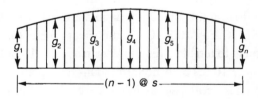

Fig. 3-35.   Simpson's rule

STRUCTURAL STEEL DESIGN

Elastic Design of Rolled Steel Girders

To ensure satisfactory performance under service load conditions, deflections due to live load plus impact are limited by AASHTO Section 10.6 to

$$\delta_{max} = L/800$$

where

$$L = \text{span length}$$

To provide additional comfort for pedestrians, the deflections of bridge structures which are also used by pedestrians are limited to

$$\delta_{max} = L/1000.$$

To achieve these limits, AASHTO Section 10.5 restricts the depth to span ratio of a steel girder to

$$d/L \geq 1/25.$$

The same limit applies to the overall depth of a composite section with the steel girder restricted to

$$d/L \geq 1/30.$$

The allowable bending stress in adequately braced steel girders is specified by AASHTO Section 10.32 as

$$F_b = 0.55F_y$$

which, for $F_y = 36$ kips per square inch, reduces to

$$F_b = 20 \text{ kips per square inch.}$$

Vehicular loading produces stress variations in the structure, which may give rise to premature fatigue failure of the structure. The allowable stress range is specified by AASHTO Section 10.3, and requires an estimate of the total number of stress cycles, the connection details and their location, and the redundancy of the structure. The number of stress cycles is determined from AASHTO Table 10.3.2A, and is based on an estimate of the average daily truck traffic. For an average daily truck traffic exceeding 2500, the number of stress cycles is given as 2×10^6 for truck loading. Connection details are listed in AASHTO Table 10.3.1B and are illustrated in AASHTO Fig. 10.3.1C. A stud shear connector, attached to the flange of a steel girder which is subject to tensile stress, is shown as item 18 in Fig. 10.3.1C and is listed as stress category C in Table 10.3.1B. A nonredundant load path structure is one in which a single fatigue failure can cause collapse, as in the case of the failure of one girder of a two girder bridge.

The bottom flange of the steel girder is subjected to tensile stress, due to dead loads and live load plus impact, and also due to lateral wind load. To control lateral forces, AASHTO Section 10.20 requires the provision of diaphragms at a maximum spacing of 25 feet. In accordance with AASHTO Sections 3.15 and 10.21 a horizontal wind pressure of 50 pounds per square foot is applied to the area of the superstructure in elevation to provide the total lateral force which shall not be less than 300 pounds per foot. Half of this force is applied to the bottom flange and the resultant stresses are given by AASHTO Equation (10-5) as

$$f_W = Rf_{cb}$$

where

$R = (0.2272L - 11)S_d^{-2/3} \ldots$ without lateral bracing

$R = (0.059L - 0.64)S_d^{-1/2} \ldots$ with lateral bracing

$f_{cb} = 72M_{cb}/t_f b_f^2$ pounds per square inch

$M_{cb} = 0.08WS_d^2$ pounds feet

L = span length

S_d = diaphragm spacing

t_f = flange thickness

b_f = flange width

W = wind load per linear foot of bottom flange

In accordance with AASHTO Section 3.22 the stresses due to live load, dead load and wind load are combined in Group II and Group III as

$$f_D + f_W \le 1.25F_b$$
$$f_D + f_{(L+I)} + 0.3f_W \le 1.25F_b$$

where,

f_D = dead load stress

$f_{(L+I)}$ = stress due to live load plus impact.

The allowable shear stress is specified by AASHTO Section 10.32 as

$F_v = 0.33F_y$

which for $F_y = 36$ kips per square inch, reduces to

$F_v = 12$ kips per square inch.

The gross web area is assumed to resist the total applied shear force and the shear stress is given by

$$f_v = V/dt_w$$

where

V = applied shear force

d = overall girder depth

t_w = web thickness.

When the shear stress in the web at the girder bearings exceeds 75 percent of the allowable shear stress, AASHTO Section 10.33 requires the provision of bearing stiffeners. The stiffeners shall extend for the full height of the web, and provide close bearing on, or be groove welded to, the flange transmitting the reaction. The stiffener plates are placed on both sides of the web and are designed as an axially loaded cruciform column with a length of web, shown in Fig. 3-36, contributing to the section properties of the column. The stiffener plates shall extend approximately to the edge of the flanges and the limiting thickness of the stiffener plates is given by AASHTO Equation (10-33) as

$$t_s \ge b' \sqrt{F_y}/2180$$

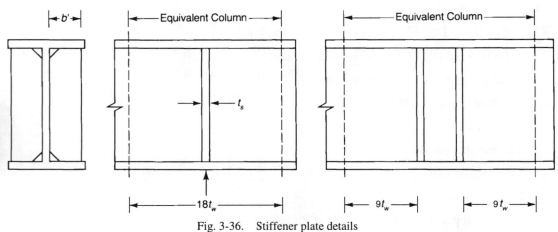

Fig. 3-36. Stiffener plate details

where

b' = stiffener plate width

The allowable bearing stress on the area of the stiffener plate in contact with the loaded flange is given by AASHTO Section 10.32 as

$$F_p = 0.80F_y$$

which, for $F_y = 36$ kips per square inch, reduces to $F_p = 29$ kips per square inch.

Load factor design may also be employed for the design of steel girders and details of the procedure is given in AASHTO Sections 10.42 to 10.60.

Example 8

Fig. 3-37 shows part of the cross section of a composite rolled steel girder bridge superstructure. Diaphragms are located at a spacing of 25 feet, and the girder is of grade A36 steel. Bending stress in the bottom flange due to dead load is 9 kips per square inch, and due to HS20 live load plus impact is 10 kips per square inch. The superstructure is simply supported with a span of 65 feet. Determine if bottom flange lateral bracing is required.

Solution

The height of the superstructure elevation is 5.75 feet and the wind pressure, in accordance with AASHTO Section 3.15, is 50 pounds per square foot. Hence, the total lateral wind force is

$W_T = 50 \times 5.75/1000$

$\quad = 0.29$ kips per foot.

The minimum specified wind force is

$W_{min} = 0.30$ kips per foot . . . governs.

Hence, the wind load per linear foot of bottom flange is

$W = W_{min}/2$

$\quad = 0.30/2$

$\quad = 0.15$ kips per foot.

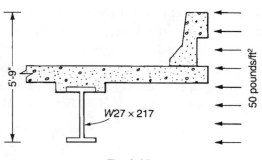

Fig. 3-37

The bending moment, due to this force, on the bottom flange is

$$M_{cb} = 0.08WS_d^2$$
$$= 0.08 \times 0.15 \times 25^2$$
$$= 7.5 \text{ kip feet}$$

The stress in the bottom flange, due to this moment is

$$f_{cb} = 72M_{cb}/t_f b_f^2$$
$$= 72 \times 7.5/(0.83 \times 14.12^2)$$
$$= 3.26 \text{ kips per square inch.}$$

The factor R is given by AASHTO Equation (10-6), when no bottom lateral bracing is provided, as

$$R = (0.2272L - 11)S_d^{-2/3}$$
$$= (0.2272 \times 65 - 11)\, 25^{-2/3}$$
$$= 0.44.$$

Hence, the relevant wind load stress is

$$f_W = Rf_{cb}$$
$$= 0.44 \times 3.26$$
$$= 1.44 \text{ kips per square inch.}$$

In accordance with AASHTO Section 3.22, the Group II stress is given by

$$f_{II} = f_D + f_W$$
$$= 9 + 1.44$$
$$= 10.44 \text{ kips per square inch}$$
$$< 1.25 F_b \ldots \text{satisfactory.}$$

The Group III stress is given by

$$f_{III} = f_D + f_{(L+I)} + 0.3f_W$$
$$= 9 + 10 + 0.3 \times 1.44$$
$$= 19.43 \text{ kips per square inch}$$
$$< 1.25F_b \ldots \text{satisfactory}$$

Hence, bottom flange lateral bracing is not required.

Elastic Design of Steel Plate Girders

To prevent instability of the plate girder during fabrication and erection, limiting dimensions are specified. The limiting width to thickness ratio of the compression flange is given by AASHTO Equation (10-19) as

$$b/t = 3250/\sqrt{f_b}$$
$$\leq 24$$

where

f_b = calculated compressive bending stress in the flange.

For F_y = 36 kips per square inch and $f_b = 0.55F_y$, this reduces to $b/t = 23$.

The limiting thickness of the web plate without longitudinal stiffeners is given by AASHTO Equation (10-23) as

$$t_w = D \sqrt{f_b}/23{,}000$$
$$\geq D/170$$

where

D = depth of web

For F_y = 36 kips per square inch and $f_b = 0.55F_y$, this reduces to

$t_w = D/165$

Provided that the web thickness is $t_w \geq D/150$ and the calculated shear stress in the web is $f_v \leq 7.33 \times 10^7/(D/t_w)^2 \leq F_y/3$, transverse intermediate stiffeners may be omitted.

For more slender webs, intermediate stiffeners are required in order to produce tension field action, with a spacing, in accordance with AASHTO Section 10.34.4, of

$$d_o \leq 3D$$

and the allowable shear stress is limited to

$$F_v = F_y[C + 0.87(1 - C)/\sqrt{1+(d_o/D)^2}]/3$$

where

$C = 6000 \sqrt{k}/\sqrt{F_y}(D/t_w)$

when, $6000\sqrt{k}/\sqrt{F_y} \leq (D/t_w) \leq 7500\sqrt{k}/\sqrt{F_y}$

and, $C = 4.5 \times 10^7 k/F_y(D/t_w)^2$

when, $D/t_w > 7500\sqrt{k}/\sqrt{F_y}$

and, $C = 1.0$

when $D/t_w < 6000\sqrt{k}/\sqrt{F_y}$

and, $k = 5 + 5/(d_o/D)^2$

When the shear stress exceeds 60 percent of this value of F_v, the bending stress in the web is limited by AASHTO Equation (10-30) to

$$F_s = F_y(0.754 - 0.34 f_v/F_v)$$

When, in accordance with AASHTO Section 10.34.4.5,

$t_w \geq D/78 \ldots$ for F_y = 36 kips per square inch

and $f_v \leq F_v$

transverse intermediate stiffeners may be omitted.

No tension field action is possible in the end panel adjacent to the simply supported end of a girder, and the length of this panel is limited to

$$d_o \le 1.5D$$

and the allowable shearing stress is given by Equation (10-29) as

$$F_v = CF_y/3$$
$$\le F_y/3.$$

Intermediate stiffeners are normally provided on one side only of the web and are welded to the compression flange and stopped short of the tension flange. The minimum moment of inertia of the stiffener is given by AASHTO Equation (10-31) as

$$I_{st} = d_o t_w^3 J$$

where

$$J = 2.5(D/d_o)^2 - 2$$
$$\ge 0.5$$

The minimum required area of the stiffener is given by AASHTO Equation (10-32a) as

$$A_{st} = 0.15 YBD t_w (1 - C)(f_v/F_v) - 18Y t_w^2$$

where

Y = ratio of yield stress of web steel to stiffener steel

$B = 1.0$ for a pair of stiffeners

$B = 1.8$ for a single angle stiffener

$B = 2.4$ for a single plate stiffener

The width of the stiffener is given by AASHTO Section 10.34.4.10 as

$$b' \ge d/30 + 2 \text{ inches}$$
$$\ge b_f/4$$

where

d = overall depth of girder

b_f = girder flange width.

The thickness of the stiffener is

$t_s \ge b'/16$

The stiffener welding, in accordance with AASHTO Section 10.34.4.9 shall terminate a distance from the toe of the web-to-flange weld between $4t_w$ and $6t_w$.

Composite Girders

A composite girder as shown in Fig. 3-38 consists of a concrete flange attached to a steel girder with shear connectors. The shear connectors provide a mechanical anchorage between the flange and the girder to transfer the horizontal shear between the two materials, without excessive slip at the interface, to ensure an integral unit. The horizontal shear is caused by dead load, live load, shrinkage of the flange concrete,[31] and differential temperature effects.[31,32]

Vehicular loading on the structure produces a variation of the shear force applied to a connector. Since this may cause premature fatigue failure, the design criterion for a connector

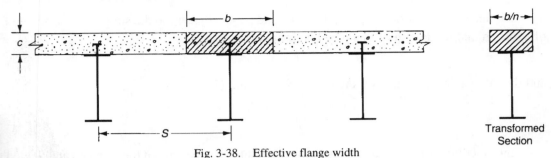

Fig. 3-38.   Effective flange width

Composite bridges may be constructed either with or without temporary props under the steel girder. In the unpropped method, the steel girder is designed to support its self weight, the weight of the cast-in-place flange concrete and the weight of all necessary formwork. Additional superimposed dead load, imposed after the concrete attains 75 percent of its 28-day compressive strength, and live load plus impact are supported by the composite section. In propped construction the steel girder supports only its self weight, and all additional loads, imposed after the props are removed, are supported by the composite section. The effect of each load is determined using the effective section properties at the stage when each load is applied, and the total effect obtained by superposition.

Because of the relatively large width of the concrete flange and the effects of shear lag, the resulting stress and strain distribution across the composite section is nonuniform. As shown in Fig. 3-38, to compensate for this effect, the actual width of flange S, between girders, is reduced to an effective width b, which is given by AASHTO Section 10.38.3 as the minimum of

(i) $b = L/4$

(ii) $b = S$

(iii) $b = 12c$

For stress computations, the composite section properties are based on the transformed section, as shown in Fig. 3-38, which is derived from the relevant modular ratio of the steel and concrete

$$n = E/E_c$$

where

E = modulus of elasticity of steel

E_c = modulus of elasticity of concrete

For short term loads and a concrete strength of 2900 to 3500 pounds per square inch, AASHTO Section 10.38.1 specifies a modular ratio of

$$n = 9$$

Under sustained load, the modular ratio increases due to creep effects in the concrete flange and the value specified is three times the short duration modulus. Hence, in determining stresses in a composite section, two transformed sections are utilized, one for loads of short duration such as live load plus impact, and one for sustained loads such as dead loads.

A continuous composite structure may be designed by assuming that negative moments at interior supports are resisted by the steel girder only. Shear connectors are not required, for this situation, over negative moment regions of the span. Alternatively, it may be assumed that negative moments at supports are resisted by a composite section consisting of the steel girder and steel reinforcement in the concrete flange. Shear connectors are then required over the negative moment regions of the span.

In order to achieve full composite action between the concrete flange and the steel girder, shear connectors must be designed to resist the fluctuation in shear at the interface due to live load plus impact. The range of horizontal shear at the interface is given by AASHTO Equation (10-58) as

$$S_r = V_r Q / I$$

where

V_r = range of shear force due to live load plus impact

Q = moment of transformed compressive concrete area about neutral axis

I = moment of inertia of transformed composite section

The allowable range of shear for a welded stud is given by Equation (10-60) as

$$Z_r = \alpha d^2 \dots \text{ for } H/d \geq 4$$

where

d = stud diameter

H = stud height

$\alpha = 13{,}000$ for 1×10^5 cycles

$\alpha = 10{,}600$ for 5×10^5 cycles

$\alpha = 7850$ for 2×10^6 cycles

$\alpha = 5500$ for over 2×10^6 cycles.

The required connector spacing is

$$p = \Sigma Z_r / S_r$$
$$\leq 24 \text{ inches}$$

where

ΣZ_r = total allowable capacity of all connectors at one transverse section.

To ensure integrity of the composite section at ultimate loads, the required number of connectors on each side of the point of maximum moment is given by AASHTO Equation (10-61) as

$$N_1 = P / \phi S_u$$

where

P = total shear force at interface at ultimate limit state

ϕ = reduction factor = 0.85

S_u = ultimate strength of shear connector

As shown in Fig. 3-39, the magnitude of the total shear force at the interface depends on the location of the neutral axis at ultimate moment and is given by the lesser of

$P = A_s F_y \dots$ neutral axis above interface

$P = 0.85 f_c' bc \dots$ neutral axis below interface

where

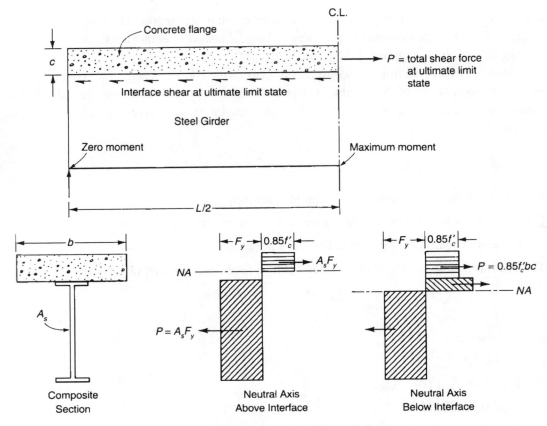

Fig. 3-39. Horizontal shear at ultimate limit state

A_s = area of steel girder

F_y = yield stress of steel

f_c' = 28-day compressive strength of flange concrete

b = effective flange width

c = flange depth

The ultimate strength of a welded stud is given by AASHTO Equation (10-67) as

$$S_u = 0.4d^2 \sqrt{f_c' E_c}$$

where

E_c = modulus of elasticity of concrete

$= w^{3/2}\, 33 \sqrt{f_c'}$

w = concrete density.

Flexural Compression

Due to the lateral instability of the compression flange, as the unsupported length l of the compression flange increases the allowable flexural compressive stress reduces and is given by AASHTO Table 10.32.1A as

$$F_b = (50 \times 10^6 C_b I_{yc}/l S_{xc}) \sqrt{0.772 J / I_{yc} + 9.87(d/l)^2}$$

where

I_{yc}= moment of inertia compression flange about a vertical axis through the web

S_{xc} = section modulus with respect to the compression flange

$J = [(bt^3)_c + (bt^3)_t + Dt_w^3]/3$

b = flange width of compression or tension flanges

t = thickness of compression or tension flanges

$C_b = 1.75 + 1.05(M_1/M_2) + 0.3(M_1/M_2)^2 \le 2.3$

M_1 = smaller moment at the end of the unsupported length

M_2 = larger moment at the end of the unsupported length

M_1/M_2 = positive for reverse curvature

When F_y = 36 kips per square inch, the allowable flexural compressive stress is given by

$F_b \le 20$ kips per square inch

**Bridge
Structures**

Compression Members

The allowable stress in an axially loaded compression member is defined in AASHTO Table 10.32.1A in terms of the slenderness ratio

$$KL/r$$

where

K = effective length factor

L = unbraced length of the column

r = the governing radius of gyration.

Fig. 2-32 of this book details the effective length factor applicable to different end conditions. Fig. 2-33 of this book presents an alignment chart for a column framing into beams at either end and subjected to sidesway.

The failure of a short, stocky strut occurs at the squash load because of yielding in compression. A factor of safety of 2.12 is adopted and the allowable stress for a slenderness ratio of zero is

$$F_a = F_y/2.12$$

which, for F_y = 36 kip per square inch, reduces to

F_a = 17 kips per square inch.

The critical slenderness ratio is

$C_c = \sqrt{2\pi E/F_y}$

= 126 for F_y = 36 kips per square inch. At this point the allowable stress is

$F_a = F_y/(2 \times 2.12)$

which, for F_y = 36 kips per square inch, reduces to

F_a = 8.5 kips per square inch.

For larger values of the slenderness ratio, the Euler elastic critical load is assumed to control and the allowable stress is given by

$F_a = \pi^2 E/2.12(KL/r)^2$

$= 135,009/(KL/r)^2$ kips per square inch

For values of the slenderness ratio between zero and the critical slenderness ratio, the allowable stress is given by the parabolic expression

$F_a = F_y[1 - (KL/r)^2F_y/4\pi^2E]/2.12$

which, for $F_y = 36$ kips per square inch, reduces to

$F_a = 17 - 0.53(KL/r)^2/1000$ kips per square inch

Combined Compression and Flexure

The flexural capacity of a member reduces in the presence of axial load, and to limit the stress at the points of support, AASHTO Equation (10-42) utilizes the interaction expression

$$f_a/0.472F_y + f_{bx}/F_{bx} + f_{by}/F_{by} \leq 1.0$$

Within the length of a member, the limiting stress is given by AASHTO Equation (10-41) as

$$f_a/F_a + C_{mx}f_{bx}/F_{bx}(1 - f_a/F'_{ex}) + C_{my}f_{by}/F_{by}(1 - f_a/F'_{ey}) \leq 1.0$$

where

F'_e = factored Euler critical stress

$\quad = \pi^2E/2.12(K_bl_b/r_b)^2$

l_b = actual unbraced length in the plane of bending

r_b = corresponding radius of gyration

K_b = corresponding effective length factor

C_m = reduction factor defined in Fig. 2-38 of this book

2.12 = factor of safety

Connections

Field splices are used in the construction of long span bridges to enable the girders to be handled and transported in convenient lengths. Because of the difficulties inherent in welding large girders in the field, field splices are usually bolted. In accordance with AASHTO Section 10.32.3, low carbon steel bolts type A307 may not be used in connections subjected to fatigue, and bearing-type high-strength bolts are undesirable due to the additional deflection produced by slip at the connections. Hence, high-strength slip-critical bolts are preferred.

The two types of girder splices utilized are shown in Fig. 3-40. The splice may consist of an outer flange cover plate on each flange, plus a web splice plate on each side

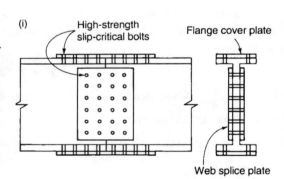
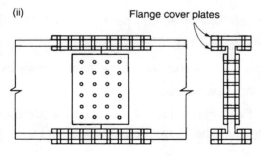

Fig. 3-40. Girder splice details

of the web as shown at (i). Alternatively, one outer flange cover plate plus two inner flange cover plates, on each flange, may be used as shown at (ii). At least two rows of bolts must be provided on each side of the joint, in the web splice plate which must extend the full depth of the girder between flanges.

The splice components must be capable of carrying the applied forces at the joint, as shown in Fig. 3-41. The web splice plate is designed for that portion of the bending moment resisted by the girder web, plus the torsional moment induced by the eccentricity of the applied shear force acting at the centroid of the bolt group on each side of the joint. In addition, the web splice plate resists the total applied shear force at the joint. The flange cover plate resists that portion of the applied moment which is not resisted by the web.

Thus, the portion of the applied moment resisted by the girder web is

$$M_{w1} = MS_w/S_G$$

where

$$S_w = t_w D^2/6$$

S_G = total section modulus of the girder.

The torsional moment induced by the connector eccentricity is given by

$$M_{w2} = eV$$

The total web splice plate design moment is, then,

$$M_w = M_{w1} + M_{w2}$$
$$= MS_w/S_G + eV$$

The total web splice plate design shear is

$$V_w = V$$

The total flange cover plate design moment is

$$M_f = M - M_{w2}$$
$$= M(1 - S_w/S_G)$$

In accordance with AASHTO Section 10.18.1, the splice components must be designed for the greater value given by

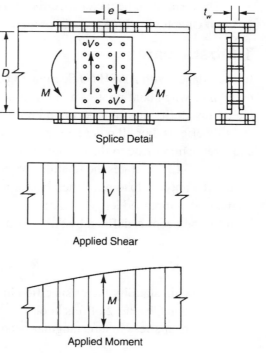

Splice Detail

Applied Shear

Applied Moment

Fig. 3-41.   Applied forces at the splice

(i) 75 percent of the allowable capacity of the girder section.

(ii) The average of the calculated design force in the component and the allowable capacity of the girder section

The allowable shear capacity of the girder web is

$$V_G = F_v A_w$$
$$= 12Dt_w \ldots \text{for } F_y = 36 \text{ kips per square inch}$$

The allowable moment capacity of the girder web is

$$M_{Gw} = F_b S_w$$
$$= 3.33 t_w D^2 \ldots \text{ for } F_y = 36 \text{ kips per square inch}$$

The allowable moment capacity of the girder flange is

$$M_{Gf} = F_b S_G - M_{Gw}$$
$$= 20(S_G - S_w) \ldots \text{ for } F_y = 36 \text{ kips per square inch.}$$

TIMBER STRUCTURES

Allowable Stresses

The allowable stresses for visually graded sawn lumber for normal duration of loading, used under dry conditions, are given in AASHTO Table 13.5.1A. Reduction factors are given for service conditions in which the moisture content exceeds 19 percent. These are identical with the factors given in Chapter 2 of this book.

The allowable stresses for glued laminated timber are given in AASHTO Tables 13.5.3A and 13.5.3B. Reduction factors are given for service conditions in which the moisture content exceeds 16 percent. These are similar to the factors given in Chapter 2 of this book.

Load duration factors are given in AASHTO Section 13.5.5 and these are similar to the factors given in Chapter 2 of this book.

The bearing adjustment factor specified in AASHTO Section 13.6 are identical to the factors given in Chapter 2.

The size factor is specified in AASHTO Section 13.6.4.2 and this is identical to the factor given in Chapter 2.

The beam volume factor specified in AASHTO Section 13.6.4.3 is identical to the factor given in Chapter 2 with $K_L = 1.0$ for all configurations and loading conditions.

Timber Beams

In accordance with AASHTO Section 13.6.4.4 when the depth of a beam does not exceed its breadth or when continuous lateral restraint is provided to the compression edge of a beam, lateral instability does not occur and allowable bending stresses require no reduction for instability effects. For all other situations, lateral instability must be investigated and the applicable reduction made to the bending stress.

Fig. 3-42 illustrates the relationship between the effective length l_e, the unbraced length l_u, and the applied loading. When lateral restraint is provided at intermediate points along a member, the unbraced length is defined as the distance between such points. The slenderness ratio for the beam is defined by AASHTO Equation (13-8) as

$$R_B = \sqrt{l_e d / b^2} \leq 50$$

The beam stability factor specified in AASHTO Section 13.6.4.4.5 is identical with the factor given in Chapter 2 of this book.

The shear stress in a rectangular beam is given by AASHTO Equation (13-9) as

$$f_v = 1.5 V / bd$$

For vehicle live loads, the governing shear force V is defined as the value occurring at a distance from the support given by the minimum of

(i) 3d

(ii) $L/4$

where, b = beam width

d = beam depth

L = span length.

In accordance with AASHTO Section 13.6.5.2 the governing shear force is determined from

$$V = (0.60V_L + V_D)/2$$

where

V_L = undistributed shear due to one line of wheel loads located over the beam

V_D = shear due to the standard truck load distributed laterally as specified for moment in AASHTO Section 3.23.

As specified, in AASHTO Section 3.8, impact is not included in timber structures due to their damping characteristics.

For a beam which is notched, the design limitations and requirements specified in AASHTO Section 13.6.2 are identical with those given in Chapter 2 of this book.

Timber Columns

The design of timber columns specified in AASHTO Section 13.7 is identical with design method given in Chapter 2 of this book.

Configuration   Effective Length, l_e

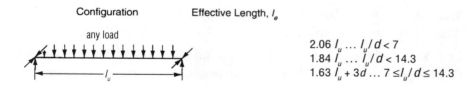

$2.06\ l_u \ldots l_u/d < 7$
$1.84\ l_u \ldots l_u/d < 14.3$
$1.63\ l_u + 3d \ldots 7 \leq l_u/d \leq 14.3$

Fig. 3-42. Beam effective length factors

References

1. American Association of State Highway and Transportation Officials. *Standard Specifications for Highway Bridges*, Sixteenth Edition, as amended by the 1997 interim revisions. Washington, D.C. 1996.

2. Williams, A. *The Analysis of Indeterminate Structures*. Macmillan, London, 1967.

3. American Institute of Steel Construction. *Moments, Shears and Reactions: Continuous Highway Bridge Tables*. Chicago, IL, 1959.

4. Jenkins, W. M. Influence Line Computations for Structures with Members of varying Flexural Rigidity Using the Electronic Digital Computer. *Structural Engineer*. Volume 39, September 1961.

5. Wang, C. K. Matrix Analysis of Statically Indeterminate Trusses. *Proceedings American Society of Civil Engineers*. Volume 85 (ST4), April 1959.

6. Portland Cement Association. *Influence Lines Drawn as Deflection Curves*. Skokie IL, 1948.

7. Thadani, B. N. Distribution of Deformation Method for the Construction of influence Lines. *Civil Engineering and Public Works Review*. Volume 51, June 1956.

8. Lee, S. L. and Patel, P. C. The Bar-Chain Method of Analyzing Truss Deformations. *Proceedings American Society of Civil Engineers*. Volume 86 (ST3), May 1960.

9. Williams, A. The Determination of Influence Lines for Bridge Decks Monolithic with their Piers. *Structural Engineer*. Volume 42, May 1964.

10. Morice, P. B. and Little, G. *The Analysis of Right Bridge Decks Subjected to Abnormal Loading*. Cement and Concrete Association, London, 1956.

11. West, R. *Recommendations on the Use of Grillage Analysis for Slab and Pseudo-Alab Bridge Decks*. Cement and Concrete Association, London, 1973.

12. Loo, Y. C. and Cusens, A. R. A Refined Finite Strip Method for the Analysis of Orthotropic plates. *Proceedings Institution of Civil Engineers*. Volume 48, January 1971.

13. Davis, J. D., Somerville, I. J. and Zienkiewicz, O. C. Analysis of Various Types of Bridges by the Finite Element Method. *Proceedings of the Conference on Developments in Bridge Design and Construction, Cardiff, March 1971*. Crosby Lockwood, London, 1972.

14. Westergaard, H. M. Computation of Stresses in Bridge Slabs Due to Wheel Loads. *Public Roads*. March, 1930.

15. Portland Cement Association. *Handbook of Frame Constants*. Skokie, IL, 1948.

16. Lee, S. L. The Conjugate Frame Method and its Application in the Elastic and Plastic Theory of Structures. *Journal Franklin Institute*. Volume 266, September 1958.

17. American Association of State Highway and Transportation Officials. *Standard Specifications for Highway Bridges. Sixteenth Edition: Division I-A Seismic Design*. Washington, D.C. 1996.

18. Building Science Safety Council. *NEHRP Recommended Provisions for the Development of Seismic Regulations for New Buildings: Part 2, Commentary.* Washington, D.C. 1997.

19. Paz, M. *Structural Dynamics.* Van Nostrand Reinhold, New York, 1991.

20. Federal Highway Administration. *Seismic Design and Retrofit Manual for Highway Bridges.* Washington, D.C. 1987.

21. Hewlett Packard Company. *HP-48G Calculator Reference Manual.* Corvallis, OR 1994.

22. Portland Cement Association. *Notes on ACI 318-95: Building Code Requirements for Reinforced Concrete.* Skokie, IL, 1995.

23. American Concrete Institute. *Building Code Requirements and Commentary for Reinforced Concrete (ACI 318-95).* Detroit, MI, 1996.

24. American Concrete Institute. *Analysis and Design of Reinforced Concrete Bridge Structures (ACI 343R-88).* Detroit, MI, 1988.

25. Reynolds, C. E. and Steedman, J. C. *Reinforced Concrete Designers Handbook.* Cement and Concrete Association, London, 1981.

26. Kajfasz, S. Somerville, G. and Rowe, R. E. *An Investigation of the Behavior of Composite Beams.* Cement and Concrete Association, London, 1963.

27. Freyermuth, C. L. *Design of Continuous Highway Bridges with Precast, Prestressed Concrete Girders.* Portland Cement Association. Skokie, IL, 1969.

28. Freyermuth, C. L. and Shoolbred, R. A. *Post-Tensioned, Prestressed Concrete.* Portland Cement Association. Skokie, IL, 1967.

29. The Concrete Society. *Post-Tensioned Flat-Slab Design Handbook.* London, 1984.

30. American Association of State Highway and Transportation Officials. *Guide Specification for Alternate Load Factor Design Procedures for Steel Beam Bridges Using Braced Compact Sections.* Washington, D.C. 1986.

31. Nash, G.F.J. *Steel Bridge Design Guide: Composite Universal Beam Simply Supported Span.* Constructional Steel Research and Development Organization. Croydon, 1984.

32. Knowles, P. R. *Simply Supported Composite Plate Girder Highway Bridge.* Constructional Steel Research and Development Organization. Croydon, 1976.

Bridge Structures

Problems and Solutions

3-1. The two-span bridge with a single central column support shown in Fig. 3-1 is located in the vicinity of San Diego on an important strategic route. The soil profile at the site consists of a 35-foot layer of soft-to-medium clay.

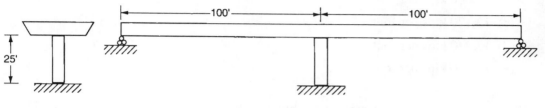

Fig. 3-1

a) The applicable acceleration coefficient, to be used for seismic design, is most nearly:

    (a) 0.40

    (b) 0.30

    (c) 0.20

    (d) 0.10

    (e) 0.05

b) What importance classification should be assigned to the structure?

    (a) I

    (b) II

c) What seismic performance category should be assigned to the structure?

    (a) A

    (b) B

    (c) C

    (d) D

d) The value of the site coefficient is most nearly:

    (a) 1.0

    (b) 1.1

    (c) 1.2

    (d) 1.3

    (e) 1.5

e) What analysis procedure should be used?

    (a) Procedure 1 or 2

    (b) Procedure 3

f) The column has a moment of inertia of 40 feet^4 and a modulus of elasticity of 432,000 kips per square foot, and may be assumed fixed at the top and bottom. The stiffness of the column is most nearly:

**Bridge
Structures**

    (a) 12,970 kips per foot

    (b) 13,070 kips per foot

    (c) 13,170 kips per foot

    (d) 13,270 kips per foot

    (e) 13,370 kips per foot

g) The weight of the superstructure and tributary substructure has a constant value of 7 kips per foot. The fundamental period of the bridge in the longitudinal direction is most nearly:

    (a) 0.34 seconds

    (b) 0.35 seconds

    (c) 0.36 seconds

    (d) 0.37 seconds

    (e) 0.38 seconds

h) The value of the elastic seismic response factor is most nearly:

    (a) 1.0

    (b) 1.1

    (c) 1.2

    (d) 1.3

    (e) 1.4

i) The elastic seismic moment in the column due to the longitudinal seismic force is most nearly:

    (a) 17,000 kip feet

    (b) 17,500 kip feet

    (c) 18,000 kip feet

    (d) 18,500 kip feet

    (e) 19,000 kip feet

j) The reduced design moment in the column is most nearly:

    (a) 5200 kip feet

    (b) 5500 kip feet

    (c) 5800 kip feet

    (d) 6000 kip feet

    (e) 6400 kip feet

Solution

a) From AASHTOSD Section 3.2, the applicable acceleration coefficient for the San Diego area is (a) = 0.4. The answer is (a).

b) From AASHTOSD Section 3.3, for a bridge located on an important strategic route, the importance classification is IC = I. The answer is (a).

c) From Table 3-1a, for a value of the acceleration coefficient exceeding 0.29 and an importance classification of *I,* the relevant seismic performance category is D (SPC = D). The answer is (d).

Table 3-1a. Seismic performance category

Acceleration Coefficient	Essential Bridges (Importance Classification I)	Other Bridges (Importance Classification II)
$A \leq 0.09$	A	A
$0.09 < A \leq 0.19$	B	B
$0.19 < A \leq 0.29$	C	C
$0.29 < A$	D	C

d) From AASHTOSD Section 3.5, the relevant site coefficient for a soft-to-medium clay layer with a depth of 35 feet is $S = 1.5$. The answer is (e).

e) From Table 3-1b, for a regular bridge with a seismic performance category of D, the required analysis procedure is Procedure 1 or 2. The answer is (a).

Table 3-1b. Analysis procedure

SPC	Bridges with 2 or more spans	
	Regular	Irregular
A	N/A	N/A
B	1 or 2	3
C	1 or 2	3
D	1 or 2	3

f) The stiffness of a column fixed at the top and bottom is given by
$k_C = 12EI/H^3$

$= 12 \times 432,000 \times 40/25^3$

$= 13,271$ kips per foot.

The answer is (d).

g) The total weight of the superstructure and tributary substructure is
$W = wL$

$= 7 \times 200$

$= 1400$ kips.

The longitudinal stiffness of the bridge is

$k_C = 13,271/12$

$\quad = 1106$ kips per inch.

The fundamental period of the bridge in the longitudinal direction is given by

$T = 0.32\sqrt{W/k_c}$

$\quad = 0.32\sqrt{1400/1106}$

$\quad = 0.36$ seconds

The answer is (c).

h) From AASHTOSD Section 3.6, the value of the elastic seismic response coefficient is given by Formula (3-1) as

$C_s = 1.2AS/T^{2/3}$

$\quad = 1.2 \times 0.40 \times 1.5/(0.36)^{2/3}$

$\quad = 1.42$

The maximum allowable value of C_s is

$C_s = 2.5A$

$\quad = 2.5 \times 0.4$

$\quad = 1.0 \ldots$ governs

The answer is (a).

i) The total elastic seismic shear is given by

$V = WC_s$

$\quad = 1400 \times 1.0$

$\quad = 1400$ kips

The elastic moment in the column is

$M_E = VH/2$

$\quad = 1400 \times 25/2$

$\quad = 17,500$ kip feet

The answer is (b).

j) The response modification factor for a single column is given in Table 3-1c as

$R = 3$

Hence, the reduced design moment in the column is

$M_R = M_E/3$

$\quad = 17,500/3$

$\quad = 5833$ kip feet

The answer is (c).

Table 3-1c. Response modification factors

Substructure	R-factor
Wall Type Pier:	
Strong axis	2
Weak axis	3
Reinforced Concrete Pile Bents:	
Vertical Piles Only	3
One or More Batter Piles	2
Single Columns:	3
Steel or Composite Pile Bents:	
Vertical Piles Only	5
One or More Batter Piles	3
Multiple Column Bent:	5

3-2. Fig. 3-2 shows the central bent in a regular two span bridge located in the vicinity of Los Angeles on a strategic route. The 4-foot diameter columns may be considered fixed at the top and bottom, and the axial force due to dead load at the bottom of each column is 800 kips. Each column is reinforced with 24 Number 14 deformed bars Grade 60, and the concrete strength is 3250 pounds per square inch. The relevant column interaction diagram is shown in the figure.

a) Do slenderness effects have to be considered in designing the columns?

    (a) Yes

    (b) No

b) From the given axial force due to dead load and the interaction diagram provided, the maximum probable plastic hinging moment at the base of each column is most nearly:

    (a) 6370 kip feet

    (b) 6470 kip feet

    (c) 6570 kip feet

    (d) 6670 kip feet

    (e) 6770 kip feet

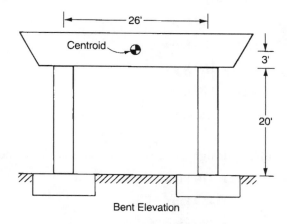

Bent Elevation

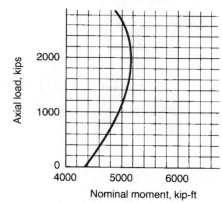

Column Interaction Diagram

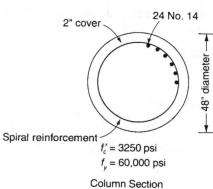

Column Section

Fig. 3-2

c) If the overstrength plastic hinge capacities at the top and bottom of the column may be assumed equal, the maximum transverse shear force developed in a column in the two column bent is most nearly:

    (a) 620 kips

    (b) 630 kips

    (c) 640 kips

    (d) 650 kips

    (e) 660 kips

d) The design shear strength provided by the concrete section outside of the end regions is most nearly:

    (a) 140 kips

    (b) 150 kips

    (c) 160 kips

    (d) 170 kips

    (e) 180 kips

e) The design shear strength required from shear reinforcement is most nearly:

    (a) 470 kips

    (b) 480 kips

    (c) 490 kips

    (d) 500 kips

    (e) 510 kips

f) The pitch required for a spiral of Number 6 reinforcement is most nearly:

    (a) 2.25 inches

    (b) 2.50 inches

    (c) 2.75 inches

    (d) 3.00 inches

    (e) 3.25 inches

g) The length of the end regions, over which special confinement reinforcement is required, is most nearly:

    (a) 40 inches

    (b) 42 inches

    (c) 44 inches

    (d) 46 inches

    (e) 48 inches

h) The minimum design shear strength provided by the concrete within the end regions of the column is most nearly:

    (a) 0 kip

    (b) 50 kips

    (c) 90 kips

    (d) 135 kips

    (e) 180 kips

i) The design shear strength required from shear reinforcement in the end regions of the left column is most nearly:

    (a) 500 kips

    (b) 530 kips

    (c) 580 kips

    (d) 620 kips

    (e) 660 kips

j) The pitch required for a spiral of Number 7 reinforcement in the end regions of the left column is most nearly:

    (a) 3.00 inches

    (b) 3.25 inches

    (c) 3.50 inches

    (d) 3.75 inches

    (e) 4.10 inches

Solution

a) From AASHTO Section 8.16.5.2, the slenderness ratio of a circular column is given by kl_u/r where

k = effective length factor

l_u = unsupported column length = 20 feet

r = radius of gyration = 0.25 × diameter = 1 foot

For an unbraced frame, with both ends of the column fixed, the effective length factor is[22]

$k = 1.0$

Hence, the slenderness ratio is

$kl_u/r = 1.0 \times 20/1.0$

$\qquad = 20$

$\qquad < 22$

Hence, the column is classified as a short column and slenderness effects may be neglected. The answer is (b).

b) From the interaction diagram, for an axial load of 800 kips, the nominal plastic hinging moment is

$M_n = 4900$ kip feet

In accordance with AASHTOSD Section 7.2.2 the overstrength plastic hinge capacity is

$M_{pr} = 1.3M_n$

$\qquad = 1.3 \times 4900$

$\qquad = 6370$ kip feet.

The answer is (a).

c) The shear forces produced in the columns by the plastic hinges, for a seismic force acting to the right, are given by

$V_{uL} = 2M_{prL}/H$

$\qquad = 2 \times 6370/20$

$\qquad = 637$ kips.

$V_{uR} = 2M_{prR}/H$

$\qquad = 2 \times 6370/20$

$\qquad = 637$ kips.

The total shear force in the bent is

$V_1 = V_{uL} + V_{uR}$

$\qquad = 637 + 637$

$\qquad = 1274$ kips.

The axial forces produced in the columns by the plastic hinges are given by

$P_{uL} = -[H_C(V_{uL} + V_{uR}) - (M_{prL} + M_{prR})]/B$

$\qquad = -[23(637 + 637) - (6370 + 6370)]/26$

$\qquad = -1127 + 490$

$\qquad = -637$ kips.

$P_{uR} = + [H_C(V_{uL} + V_{uR}) - (M_{prL} + M_{prR})]/B$

$\qquad = 1127 - 490$

$\qquad = 637$ kips.

The axial forces produced in the columns by the dead load plus the plastic hinges are given by

$P_L = P_D + P_{uL}$

$\qquad = 800 - 637$

$\qquad = 163$ kips

Bridge Structures

$$P_R = P_D + P_{uR}$$
$$= 800 + 637$$
$$= 1437 \text{ kips}$$

Using these revised axial forces, the nominal plastic hinging moments in the columns are obtained from the interaction diagram as

$M_{nL} = 4450$ kip feet

$M_{nR} = 5100$ kip feet

The corresponding overstrength plastic hinge capacities are given by

$$M_{prL} = 1.3 \times 4450$$
$$= 5785 \text{ kip feet}$$

$$M_{prR} = 1.3 \times 5100$$
$$= 6630 \text{ kip feet}$$

The shear forces produced in the columns by the revised plastic hinge capacities are given by

$$V_{uL} = 2M_{prL}/H$$
$$= 2 \times 5785/20$$
$$= 579 \text{ kip feet}$$

$$V_{uR} = 2M_{prR}/H$$
$$= 2 \times 6630/20$$
$$= 663 \text{ kip feet}$$

The total shear force in the bent is

$$V_2 = V_{uL} + V_{uR}$$
$$= 579 + 663$$
$$= 1242 \text{ kips}$$

The percentage change in total shear between Cycle 1 and Cycle 2 is

$$\Delta V = 100 \ (V_1 - V_2)/V_1$$
$$= 100 \ (1274 - 1242)/1274$$
$$= 2.5\%$$
$$< 10\%.$$

Hence, in accordance with AASHTOSD Section 7.2.2 no further iterations are necessary and the maximum probable shear force in the right column is

$V_{\max} = 663$ kips

The answer is (e).

d) In accordance with AASHTO Section 8.16.6.2 the design shear strength provided by the concrete, outside of the end regions, is given by Equations (8-46) and (8-49) as

$$\phi V_c = 2\phi bd \sqrt{f_c'}$$

where, ϕ = strength reduction factor = 0.85 from Section 8.16.1.2

b = column diameter = 48 inches

d = distance from compressive fibre to centroid of reinforcement in opposite side of member

= 37 inches

f_c' = concrete compressive strength = 3250 pounds per square inch

Hence, $\phi V_c = 2 \times 0.85 \times 48 \times 37 \times \sqrt{3250}/1000$

= 172 kips.

The answer is (d).

e) The design shear strength required from the shear reinforcement is given by AASHTO Equation (8-46) and (8-47) as

$\phi V_s = V_u - \phi V_c$

= 663 – 172

= 491 kips

$< 8\phi V_c$.

Hence, in accordance with AASHTO Section 8.16.6.3, the required shear reinforcement strength is within allowable limits. The answer is (c).

f) To satisfy the requirements for lateral reinforcement in a compression member, in accordance with AASHTO Section 8.18.2, the clear distance between spirals shall not exceed 3 inches and the minimum volumetric ratio of the spiral reinforcement to the concrete core is given by Equation (8-63) as

$\rho_s = 0.45 (A_g/A_c - 1) f_c'/f_y$

$= A_v \pi (D_c - D_s)/sA_c$

where, A_g = gross area of column = 1810 square inches

A_c = area of concrete core measured to outside of spiral = 1521 square inches

D_c = diameter of core measured to outside of spiral = 44 inches

D_s = diameter of spiral reinforcement = 0.75 inches

A_v = area of spiral reinforcement = 0.44 square inches

s = pitch of spiral reinforcement

Hence, the required pitch is given by

$s = A_v f_y \pi (D_c - D_s)/0.45 A_c f_c' (A_g/A_c - 1)$

$= 0.44 \times 60,000 \pi (44 - 0.75)/[0.45 \times 1521 \times 3250 (1810/1521 - 1)]$

= 8.5 inches

s = 3.75 inches maximum . . . governs.

To satisfy the requirements for shear strength, in accordance with AASHTO Section 8.16.6.3, the required spiral pitch is given by Equation (8-53) as

$$s = \phi A_v f_y d / \phi V_s$$

$$= 0.85 \times 2 \times 0.44 \times 60 \times 37/491$$

$$= 3.38 \text{ inches}$$

$$< 3.75 \text{ inches.}$$

The answer is (e).

g) In accordance with AASHTOSD Section 4.6.2 the length of the end regions is the larger of

(i) 18 inches

(ii) $H/6 = 20 \times 12/6 = 40$ inches

(iii) Column diameter = 48 inches . . . governs

The answer is (e).

h) The final axial forces produced in the columns by the final values of the overstrength plastic hinge moments are given by

$$P_{uL} = -[H_C(V_{uL} + V_{uR}) - (M_{prL} + M_{prR})]/B$$

$$= -[23(579 + 663) - (5785 + 6630)]/26$$

$$= -1098 + 478$$

$$= -621 \text{ kips}$$

$$P_{uR} = [H_C(V_{uL} + V_{uR}) - (M_{prL} + M_{prR})]/B$$

$$= 1098 - 478$$

$$= 621 \text{ kips}$$

The final axial forces produced in the columns by the dead load plus the plastic hinges are given by

$$P_L = P_D + P_{uL}$$

$$= 800 - 621$$

$$= 179 \text{ kips}$$

$$P_R = P_D + P_{uR}$$

$$= 800 + 621$$

$$= 1421 \text{ kips}$$

The axial force value given by

$$A_g f_c' /10 = 1810 \times 3.25/10$$

$$= 588 \text{ kips}$$

$$> P_L$$

Hence, in accordance with AASHTOSD Section 7.6.2, the design shear strength of the concrete in the end region of the left column is given by

$$\phi V_c{}' = \phi V_c P_L / 588$$

$$= 172 \times 179 / 588$$

$$= 52 \text{ kips}$$

The answer is (b).

i) For the left column, the design shear strength required from the shear reinforcement is given by AASHTO Equations (8-46) and (7-5) as

$$\phi V_s = V_{uL} - \phi V_c{}'$$

$$= 579 - 52$$

$$= 527 \text{ kips.}$$

The answer is (b).

j) The required pitch of confinement reinforcement is given by the smaller value obtained from AASHTOSD Equations (7-4) and (7-5). Then, for Number 7 reinforcement,

$$s = A_v f_y \pi (D_c - D_s) / 0.45 A_c f_c' (A_g / A_c - 1)$$

$$= 0.60 \times 60{,}000 \pi (44 - 0.875) / [0.45 \times 1521 \times 3250 \, (1810/1521 - 1)]$$

$$= 11.6 \text{ inches.}$$

or, $s = A_v f_y \pi (D_c - D_s) / 0.12 \, A_c f_c'$

$$= 0.60 \times 60{,}000 \pi (44 - 0.875) / [0.12 \times 1521 \times 3250]$$

$$= 8.22 \text{ inches.}$$

To satisfy the requirements for shear strength, in accordance with AASHTO Section 8.16.6.3, the required spiral pitch is given by Equation (8-53) as

$$s = \phi A_v f_y d / \phi V_s$$

$$= 0.85 \times 2 \times 0.60 \times 60 \times 37 / 527$$

$$= 4.30 \text{ inches.}$$

Since the clear distance between spirals shall not exceed 3 inches, in accordance with AASHTO Section 8.18.2, the maximum pitch is

$$s = 3 + D_s$$

$$= 3 + 0.875$$

$$= 3.875 \ldots \text{ governs}$$

The answer is (d).

3-3. Fig. 3-3 shows part of the superstructure of a two-lane highway bridge with an effective span of 44 feet. The precast, pretensioned girders have a 28-day compressive strength of 6000 pounds per square inch, and the final prestressing force, acting at the position shown, has a magnitude of 300 kips, all losses having occurred before the flange is cast. The area of the stress-relieved strand is two square inches, and the strand has a specified tensile strength

Superstructure Details

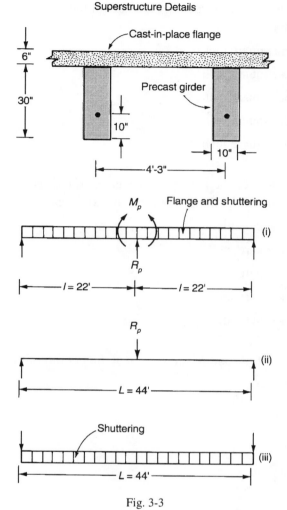

Fig. 3-3

of 270 kips per square inch. Before placing the flange shuttering and casting the flange, each girder is propped at its center with a firm, rigid prop. The shuttering is supported by the girders and weighs 30 pounds per foot run per rib, and the 28-day compressive strength of the concrete in the flange is 3000 pounds per square inch. The design load consists of HS20 loading. The effects of differential shrinkage and additional superimposed dead load, added after the flange is cast, may be ignored.

a) The stress in the bottom of the girder, at the center of the span, attributable to the final prestressing force only, is most nearly:

   (a) 1.80 kips per square inch

   (b) 1.90 kips per square inch

   (c) 2.00 kips per square inch

   (d) 2.10 kips per square inch

   (e) 2.20 kips per square inch

b) The stress in the bottom of the girder, at the center of the span, due to the self weight of the girder only, is most nearly:

    (a) −0.50 kips per square inch

    (b) −0.60 kips per square inch

    (c) −0.70 kips per square inch

    (d) −0.80 kips per square inch

    (e) −0.90 kips per square inch

**Bridge
Structures**

c) The stress in the bottom of the girder, at the center of the span, due to the weight of the shuttering plus the weight of the flange concrete, is most nearly:

    (a) 0.17 kips per square inch

    (b) 0.19 kips per square inch

    (c) 0.21 kips per square inch

    (d) 0.23 kips per square inch

    (e) 0.25 kips per square inch

d) The section modulus of the composite section at the bottom of the girder is most nearly:

    (a) 2400 inches^3

    (b) 2500 inches^3

    (c) 2600 inches^3

    (d) 2700 inches^3

    (e) 2800 inches^3

e) The stress produced in the bottom of the girder, at the center of the span, by the removal of the prop, is most nearly:

    (a) −0.25 kips per square inch

    (b) −0.35 kips per square inch

    (c) −0.45 kips per square inch

    (d) −0.55 kips per square inch

    (e) −0.65 kips per square inch

f) The stress produced in the bottom of the girder, at the center of the span, by the removal of the shuttering, is most nearly:

    (a) 0.03 kips per square inch

    (b) 0.04 kips per square inch

    (c) 0.05 kips per square inch

    (d) 0.06 kips per square inch

    (e) 0.07 kips per square inch

g) The maximum moment in the girder due to HS20 standard loading causes a stress at the bottom of the girder which is most nearly:

    (a) 0.80 kips per square inch

    (b) 0.90 kips per square inch

    (c) 1.00 kips per square inch

    (d) 1.10 kips per square inch

    (e) 1.20 kips per square inch

h) The final bottom fibre stress in the girder, at the center of the span, due to all causes, is most nearly:

    (a) 0.03 kips per square inch

    (b) 0.04 kips per square inch

    (c) 0.05 kips per square inch

    (d) 0.06 kips per square inch

    (e) 0.07 kips per square inch

i) The design flexural capacity of the composite section is most nearly:

    (a) 760 kip feet

    (b) 820 kip feet

    (c) 880 kip feet

    (d) 940 kip feet

    (e) 1000 kip feet

j) The maximum factored moment at mid span is most nearly:

    (a) 760 kip feet

    (b) 820 kip feet

    (c) 880 kip feet

    (d) 940 kip feet

    (e) 1000 kip feet

Solution

a) The relevant properties of the precast girder are

$A = 10 \times 30 = 300$ inches^2

$S_b = 10 \times 30^2/6 = 1500$ inches^3

$e = 30/2 - 10 = 5$ inches . . . lower kern position

Hence, the stress in the bottom of the girder due to the prestressing force is

$f_b = 2P_e/A$

$= 2 \times 300/300$

$= 2.0$ kips per square inch

The answer is (c).

b) The bending moment due to the girder self weight is

$M_G = 0.150 \times 300 \times 44^2 \times 12/(144 \times 8)$

$= 915$ kip inches

The stress in the bottom of the girder due to this moment is

$f_b = -M_G/S_b$

$= -915/1500$

$= -0.610$ kips per square inch

The answer is (b).

c) The weight of the shuttering is

$w_S = 0.030$ kips per foot

The weight of the flange concrete is

$w_F = 0.15 \times A_F$

$= 0.15 \times 4.25 \times 0.5$

$= 0.319$ kips per foot.

The total weight of the shuttering plus flange concrete is

$w = w_S + w_F$

$= 0.030 + 0.319$

$= 0.349$ kips per foot.

The central prop creates a two span beam with spans of 22 feet as shown at (i), with a central reaction of

$R_P = 10wl/8$

$= 10 \times 0.349 \times 22/8$

$= 9.60$ kips.

The moment at the prop due to the shuttering plus flange concrete is

$M_P = wl^2/8$

$= 1.5 \times 0.349 \times 22^2$

$= 253$ kip inches.

The stress in the bottom of the girder due to this moment is

$f_b = M_P/S_b$

$= 253/1500$

= 0.169 kips per square inch.

The answer is (a).

d) The modular ratio for the cast-in-place flange and the precast girder is given by AASHTO Section 9.16.2.1 as

$$n = E_f E_r$$
$$= \sqrt{f'_{c(\text{flange})} / f'_{c(\text{girder})}}$$
$$= \sqrt{3000 / 6000}$$
$$= 0.707.$$

The effective compression flange width is given by AASHTO Section 9.8.1 and Section 8.10.1 as the minimum of

(1) $b = L/4$

$= 44/4$

$= 11$ feet

(2) $b = b_w + 12h_f$

$= 10 + 12 \times 6$

$= 82$ inches

(3) $b = S$

$= 4.25 \times 12$

$= 51$ inches . . . governs

The transformed flange width is

$b_t = nb$

$= 0.707 \times 51$

$= 36$ inches.

The section properties of the composite section are obtained as shown in Table 3-3. Hence,

$\bar{y} = \Sigma Ay/\Sigma A$

$= 11628/516$

$= 22.5$ inches

$I_C = \Sigma I + \Sigma Ay^2 - \bar{y}^2 \Sigma A$

$= 63,836$ inches^4

Table 3-3.   Composite section properties

Part	A	y	I	Ay	Ay^2
Girder	300	15	22,500		
Flange	216	33	648		
Total	516		23,148	11,628	302,724

The section modulus at the bottom of the section is

$$S_{Cb} = I_c / \bar{y}$$
$$= 2833 \text{ inches}^3$$

The answer is (e).

e) Removal of the prop is equivalent to applying a downward load to the composite section, at mid span, equal in magnitude to the reaction in the prop, as shown at (ii). This produces the moment

$$M_R = R_p L/4$$
$$= 9.6 \times 44 \times 12/4$$
$$= 1267 \text{ kip inches.}$$

The stress in the bottom of the girder due to this moment is

$$f_{Cb} = -M_r / S_{Cb}$$
$$= -1267/2833$$
$$= -0.447 \text{ kips per square inch.}$$

The answer is (c).

f) Removal of the shuttering is equivalent to applying an upward load to the composite section equal in magnitude to the weight of the shuttering, as shown at (iii). This produces the moment

$$M_S = w_S L^2/8$$
$$= 1.5 \times 0.03 \times 44^2$$
$$= 87.12 \text{ kips inches.}$$

The stress in the bottom of the girder due to this moment is

$$f_{Cb} = M_S / S_{Cb}$$
$$= 87.12/2833$$
$$= 0.031 \text{ kips per square inch.}$$

The answer is (a).

g) The maximum moment due to standard HS20 loading in a span length of 44 feet is produced by the standard truck loading and is given by

$$M = 18L - 280 + 391/L$$
$$= 18 \times 44 - 280 + 391/44$$
$$= 521 \text{ kip feet.}$$

This moment occurs under the central axle, which is 2.33 feet from the center of the span. The impact fraction is given by AASHTO Formula (3-1) as

$$I = 50/(L + 125)$$
$$= 50/(44 + 125)$$
$$= 0.30$$

The maximum moment due to live load plus impact is

$$M_L = M(1 + 0.3)$$

$$= 521 \times 1.3$$

$$= 677 \text{ kip feet.}$$

In accordance with AASHTO Section 3.23, the fraction of the wheel live load which is distributed to an interior girder is

$$G = S/5.5 \dots \text{ for concrete slab on prestressed girders}$$

$$= 4.25/5.5$$

$$= 0.77.$$

Hence, the maximum moment due to live load plus impact distributed to the interior girder is

$$M_{max} = GM_L/2$$

$$= 0.77 \times 677/2$$

$$= 261 \text{ kip feet}$$

The stress in the bottom of the girder due to this moment is

$$f_{Cb} = -M_{max}/S_{Cb}$$

$$= -12 \times 261/2833$$

$$= -1.104 \text{ kips per square inch.}$$

The answer is (d).

h) The final bottom fibre stress in the girder is

$$f_b = 2.000 - 0.610 + 0.169 - 0.447 + 0.031 - 1.104$$

$$= 0.039 \text{ kips per square inch, compression.}$$

The answer is (b).

i) The ratio of prestressing steel is

$$p^* = A_s^*/bd$$

Assuming that the neutral axis lies within the flange

$$p^* = 2/(51 \times 26)$$

$$= 0.00151$$

The effective prestress in the tendons after all losses is

$$f_{se} = P_e/A_s^*$$

$$= 300/2$$

$$= 150 \text{ kips per square inch}$$

$$> 0.5 f_s' \dots \text{ satisfactory.}$$

The compression zone factor, for 6000 pounds per square inch concrete, given by AASHTO Section 8.16.2.7 is

$$\beta_1 = 0.85 - (6 - 4)/20$$
$$= 0.75$$

The prestressing steel factor, for stress-relieved strand, given by AASHTO Section 9.1.2 is

$$\gamma^* = 0.40$$

Hence, the stress in the bonded tendons at ultimate load is given by AASHTO Equation (9-17) as

$$f_{su}^* = f_s'(1 - \gamma^* p^* f_s'/\beta_1 f_c')$$
$$= 270[1 - 0.40 \times 0.00151 \times 270/(0.75 \times 3)]$$
$$= 250.4 \text{ kips per square inch}$$

The reinforcement index is given by AASHTO Equation (9-20) as

$$R_i = p f_{su}^* / f_c'$$
$$= 0.00151 \times 250.4/3$$
$$= 0.126$$
$$< 0.36\beta_1 \ldots \text{ satisfactory}$$

The neutral axis depth is given by AASHTO Section 9.17.2 as

$$c = A_s^* f_{sy}^* / 0.85 f_c' b$$
$$= 2 \times 250.4/(0.85 \times 3 \times 51)$$
$$= 3.85 \text{ inches}$$
$$< 6 \text{ inches}$$

Hence, the neutral axis lies within the flange.

For factory produced precast members, AASHTO Section 9.14 specifies that the strength reduction factor is

$$\phi = 1.0.$$

Hence, the flexural design capacity is given by AASHTO Equation (9-13) a

$$\phi M_n = \phi A_s^* f_{su}^* d (1 - 0.60 R_i)$$
$$= 1.0 \times 2 \times 250.4 \times 27 (1 - 0.60 \times 0.126)/12$$
$$= 1042 \text{ kip feet}$$

The answer is (e).

j) The bending moment due to the self weight of the girder is

$$M_G = 915/12$$
$$= 76 \text{ kip feet.}$$

The bending moment due to the weight of the flange concrete is

$$M_F = w_F L^2/8$$

$$= 0.319 \times 44^2/8$$

$$= 77 \text{ kip feet.}$$

The bending moment due to live load plus impact is

$$M_{max} = 261 \text{ kip feet.}$$

The factored ultimate moment is given by AASHTO Equation (3-10) as

$$M_u = \gamma[\beta_D D + \beta_L(L + I)]$$

$$= 1.3\ [1.0(76 + 77) + 1.67 \times 261]$$

$$= 766 \text{ kip feet.}$$

The answer is (a).

3-4. Fig. 3-4 shows the elevation of a two-span, post-tensioned, box section, bridge super-structure. The properties of the box section, which are constant over the whole length of the bridge, are shown on the figure. The variation of the cable eccentricity is indicated on the figure and the final prestressing force of 7000 kips may be considered constant over the whole length of the bridge.

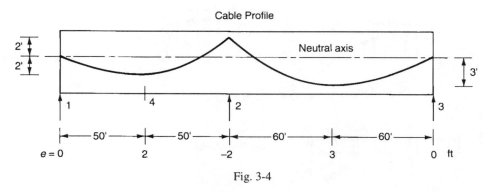

Fig. 3-4

a) The primary bending moment at Support 2, due to the final prestressing force, is most nearly:

    (a) 10,000 kip feet

    (b) 14,000 kip feet

    (c) 18,000 kip feet

    (d) 21,000 kip feet

    (e) 25,000 kip feet

b) The primary bending moment at Section 4 is most nearly:

    (a) 10,000 kip feet

    (b) 14,000 kip feet

    (c) 18,000 kip feet

    (d) 21,000 kip feet

    (e) 25,000 kip feet

c) The moment at Support 2, due to combined primary and secondary moments, is most nearly:

 (a) 10,000 kip feet

 (b) 14,000 kip feet

 (c) 18,000 kip feet

 (d) 21,000 kip feet

 (e) 25,000 kip feet

d) The secondary moment at Support 2 is most nearly:

 (a) 5400 kip feet

 (b) 7000 kip feet

 (c) 8600 kip feet

 (d) 10,800 kip feet

 (e) 13,000 kip feet

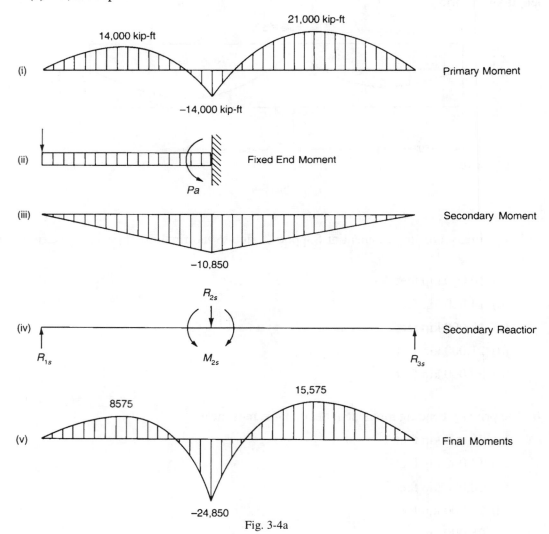

Fig. 3-4a

e) The secondary reaction at Support 1 is most nearly:

  (a) 90 kips

  (b) 110 kips

  (c) 140 kips

  (d) 170 kips

  (e) 200 kips

f) The secondary reaction at Support 3 is most nearly:

  (a) 90 kips

  (b) 110 kips

  (c) 140 kips

  (d) 170 kips

  (e) 200 kips

g) The secondary reaction at Support 2 is most nearly:

  (a) 90 kips

  (b) 110 kips

  (c) 140 kips

  (d) 170 kips

  (e) 200 kips

h) The secondary moment at Section 4 is most nearly:

  (a) 5400 kip feet

  (b) 7000 kip feet

  (c) 8600 kip feet

  (d) 10,800 kip feet

  (e) 13,000 kip feet

i) The moment at Section 4, due to combined primary and secondary moments, is most nearly:

  (a) 5400 kip feet

  (b) 7000 kip feet

  (c) 8600 kip feet

  (d) 10,800 kip feet

  (e) 13,000 kip feet

j) The stress in the bottom fibre of the section at Support 2, due to combined primary and secondary moments, is most nearly:

(a) 0.30 kips per square inch

(b) 0.45 kips per square inch

(c) 0.60 kips per square inch

(d) 0.75 kips per square inch

(e) 0.90 kips per square inch

Solution

a) The primary moment at Support 2, as shown at (i), is

$M_{2p} = Pe$

$\quad = 7000 \times (-2)$

$\quad = -14,000$ kip feet, sagging

The answer is (b).

b) The primary moment at Section 4, as shown at (i), is

$M_{4p} = Pe$

$\quad = 7000 \times 2$

$\quad = 14,000$ kip feet, hogging.

The answer is (b).

c) The drape of the parabolic cable in Span 12 is

$a = 2 + 2/2$

$\quad = 3$ feet.

The equivalent upward lateral balancing load is obtained from Figure 2-21 as

$w = 8Pa/L^2$

Allowing for the hinge at Support 1, the initial fixed-end moment at Support 2, as shown at (ii) is

$M_{21}^F = -wL^2/8$, anticlockwise

$\quad = -Pa$

$\quad = -7000 \times 3$

$\quad = -21,000$ kip feet.

The drape of the parabolic cable in Span 23 is

$a = 3 + 2/2$

$\quad = 4$ feet.

Allowing for the hinge at Support 3, the initial fixed-end moment at Support 2 is

$M_{23}^F = Pa$, clockwise

$\quad = 7000 \times 4$

$\quad = 28,000$ kip feet.

Table 3-4. Moment distribution

Joint	1	2	2	3
Member	12	21	23	32
Relative EI/L		3/100	3/120	
Distribution factors		0.55	0.45	
Fixed-end moment		−21,000	28,000	
Distribution		−3,850	−3,150	
Final moments		−24,850	24,850	

These initial fixed-end moments are distributed as shown in Table 3-4. Advantage is taken of the hinged supports to eliminate carry-over to Ends 1 and 3.

The final moment at Support 2, due to combined primary and secondary moment, is

$M_{2e} = -24{,}850$ kip feet, sagging

The answer is (e).

d) The secondary moment at Support 2, as shown at (ii), is

$M_{2s} = M_{2e} - M_{2p}$

$\quad = -24{,}850 + 14{,}000$

$\quad = -10{,}850$ kip feet, sagging

The answer is (d).

e) The secondary reaction at Support 1, as shown at (iv), is given by

$R_{1s} = M_{2s}/L_{12}$

$\quad = -10{,}850/100$

$\quad = -108.5$ kips, upward.

The answer is (b).

f) The secondary reaction at Support 3, as shown at (vi), is given by

$R_{3s} = M_{2s}/L_{23}$

$\quad = -10{,}850/120$

$\quad = -90.4$ kips, upward.

The answer is (a).

g) The secondary reaction at Support 2, as shown at (iv), is given by

$R_{2s} = -(R_{1s} + R_{3s})$

$\quad = 108.5 + 90.4$

$\quad = 198.9$ kips, downward.

The answer is (e).

h) The secondary moment at Section 4, as shown at (iii), is

$$M_{4s} = R_{1s}L_{12}/2$$

$$= M_{2s/2}$$

$$= -5425 \text{ kip feet, sagging.}$$

The answer is (a).

i) The final moment at Section 4, as shown at (v), is given by

$$M_{4e} = M_{4p} + M_{4s}$$

$$= 14{,}000 - 5425$$

$$= 8575 \text{ kip feet}$$

The answer is (c).

j) The final bottom fibre stress at Section 2 is given by

$$f_{2b} = P/A + M_{2e}/S_b$$

$$= 7000/7000 - 24{,}850 \times 12/170{,}000$$

$$= -0.754 \text{ kips per square inch, tension.}$$

The answer is (d).

3-5. Details of a polygonal two-hinged arch are given in Fig. 3-5. The relative EI values are shown ringed.

Requirements:

Determine the influence line ordinates for horizontal thrust at Hinge 1 and Hinge 2, indicating values at twenty feet centers, as unit load traverses the beam from Node 3 to Node 7.

Solution

The influence line ordinates are obtained by Maxwell's reciprocal theorem using the conjugate beam method[2,16].

The cut-back structure is produced by releasing the horizontal restraints at Hinge 1 and Hinge 2 and applying a horizontal force of 0.3 units. The corresponding supports in the conjugate structure consist of fixed-ends and the elastic load on the conjugate structure consists of the bending moment in the cut-back structure divided by the EI values, as shown in the figure. The rotation of the cut-back structure at Joint 1 is given by the reaction in the conjugate structure at Joint 1′ which is

$$R_1' = \theta_1 = (2 \times 12 \times 50/2 + 4 \times 80)/2$$

$$= 460$$

The total horizontal displacement of the cut-back structure between Support 1 and Support 2 is given by the total moment of the conjugate structure about the axis through Joints 1′ and 2′ which is

$$M_1' + M_2' = \delta_1 + \delta_2 = 2 \times 12 \times 50 \times 80/(2 \times 3) + 4 \times 80 \times 40$$

$$= 28{,}800$$

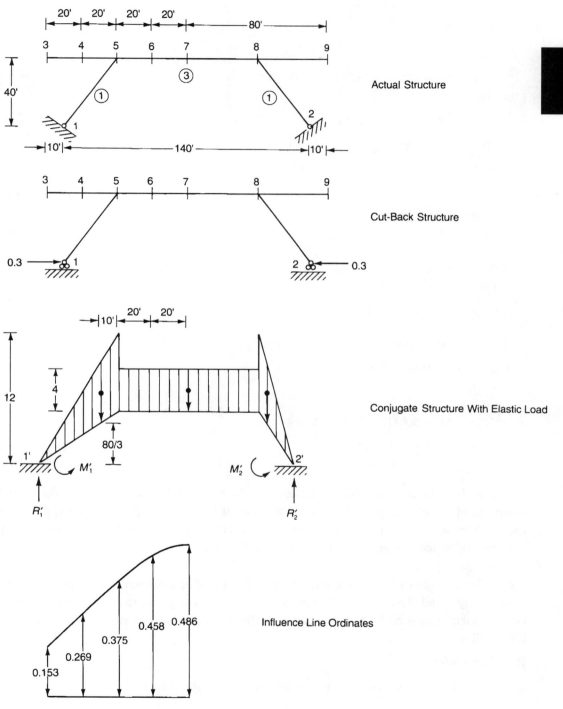

Actual Structure

Cut-Back Structure

Conjugate Structure With Elastic Load

Influence Line Ordinates

Fig. 3-5

The rotation of the cut-back structure at Node 5 is given by the shear in the conjugate structure at Node 5' which is

$$Q_5' = \theta_5 = R_1' - 12 \times 50 / 2$$

$$= 460 - 300$$

$$= 160$$

The vertical ordinates of the elastic curve due to the horizontal force of 0.3 units are given by

$$\delta_5 = M_5' = 30 R_1' - 10 \times 12 \times 50 / 2$$

$$= 30 \times 460 - 3000$$

$$= 10,800$$

$$\delta_3 = \delta_5 - 40\theta_5$$

$$= 10,800 - 40 \times 160$$

$$= 4400$$

$$\delta_4 = \delta_5 - 20\theta_5$$

$$= 10,800 - 20 \times 160$$

$$= 7600$$

$$\delta_6 = M_6' = 50 R_1' - 30 \times 12 \times 50 / 2 - 10 \times 4 \times 20$$

$$= 50 \times 460 - 9000 - 800$$

$$= 13,200$$

$$\delta_7 = M_7' = 70 R_1' - 50 \times 12 \times 50 / 2 - 20 \times 4 \times 40$$

$$= 70 \times 460 - 15,000 - 3200$$

$$= 14,000$$

Dividing the ordinates of the elastic curve by the total displacement $(\delta_1 + \delta_2)$ gives the ordinates of the elastic curve for unit displacement between Support 1 and Support 2, and these are the influence line ordinates for horizontal thrust between Hinge 1 and Hinge 2. The required influence line ordinates are shown on the Fig.

3-6. Fig. 3-6 shows the piling arrangement for a bridge abutment. The pile cap may be assumed rigid and the soil under the cap carries no load. The piles are end bearing and may be considered hinged at each end with negligible flexural stiffness. The value of *AE/L* is constant for all piles

Requirements:

Determine the axial forces in the piles due to the loading indicated.

Solution

The displacement of the cap and piles is a combination of horizontal translation, vertical translation and rotation. The solution may be obtained by means of the displacement method of analysis. Adopting matrix notation, the vector of pile cap displacements is given by

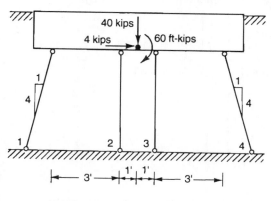

Fig. 3-6

$$[\Delta] = \begin{bmatrix} x \\ y \\ \theta \end{bmatrix}$$

The vector of pile axial forces is

$$[P] = \begin{bmatrix} P_1 \\ P_2 \\ P_3 \\ P_4 \end{bmatrix}$$

The vector of applied loads is

$$[W] = \begin{bmatrix} H \\ V \\ M \end{bmatrix} = \begin{bmatrix} 4 \\ 40 \\ 60 \end{bmatrix}$$

The diagonal matrix formed from the axial stiffnesses of the individual piles is

$$[\bar{S}] = AE/L \begin{bmatrix} 1 & 0 & 0 & 0 \\ 0 & 1 & 0 & 0 \\ 0 & 0 & 1 & 0 \\ 0 & 0 & 0 & 1 \end{bmatrix}$$

The elements of the transformation matrix are the axial deformations produced in the piles by unit value of each pile displacement imposed in turn and the transformation matrix is given by

$$[T] = \begin{bmatrix} -\frac{1}{4} & 1 & -4 \\ 0 & 1 & -1 \\ 0 & 1 & 1 \\ \frac{1}{4} & 1 & 4 \end{bmatrix}$$

The complete stiffness matrix for the whole system is

$$[S] = [T]^T [\bar{S}] [T] = AE/L \begin{bmatrix} \frac{1}{8} & 0 & 2 \\ 0 & 4 & 0 \\ 2 & 0 & 34 \end{bmatrix}$$

The vector of pile cap displacements is given by

$$[\Delta] = [S]^{-1}[W] = L/AE \begin{bmatrix} 64 \\ 10 \\ -2 \end{bmatrix}$$

The vector of pile axial forces is given by

$$[P] = [\bar{S}][T][\Delta] = \begin{bmatrix} 2 \\ 12 \\ 8 \\ 18 \end{bmatrix}$$

3-7. Fig. 3-7 shows a symmetrical, three span continuous bridge superstructure. The flexural rigidity of the superstructure varies and the relative EI values are shown. A prestressing force of 100 kips is applied to the superstructure, and this may be assumed to be constant along the whole length. The variation of the cable eccentricity is also shown.

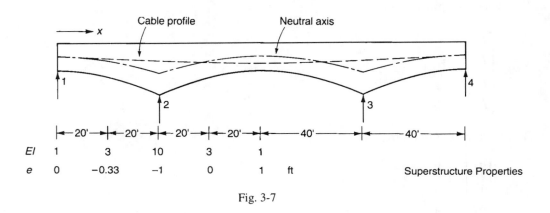

EI	1	3	10	3	1	
e	0	−0.33	−1	0	1	ft

Superstructure Properties

Fig. 3-7

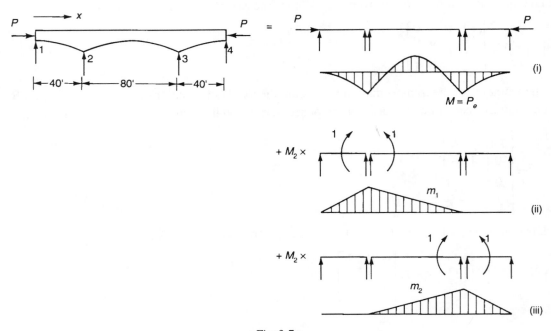

Fig. 3-7a

Requirements:

Determine the distribution of moment produced in the superstructure by the prestressing force
and secondary effects.

Solution

Due to the symmetry of the structure and loading, the reactions at Supports 2 and 3 are equal.
The moments M_2 and M_3, at these supports are taken as the redundants and releases introduced at 2 and 3 to produce the cut-back structure.

The prestressing force applied to the cut-back structure produces the distribution of
moment M shown at Fig. 3-7(a) (i). Unit value of each redundant applied in turn to the cut-
back structure produces the moments m_2 and m_3 shown at (ii) and (iii). Values of M, m_2 and
m_3 are tabulated in Table 3-7.

Table 3-7. Numerical integration

x	0	20	40	60	80	100	120
EI	1	3	10	3	1	3	10
e	0	−0.33	−1	0	1	0	0
$M = Pe$	0	−33	−100	0	100	0	−100
m_2	0	1/2	1	3/4	1/2	1/4	0
m_3	0	0	0	1/4	1/2	3/4	1
Mm_2/EI	0	−11/2	−10	0	50	0	0
m_2^2/EI	0	1/12	1/10	3/16	1/4	1/48	0
$m_2 m_3$EI	0	0	0	1/16	1/4	1/16	0
$m_2 M_2$	0	−10.1	−20.2	−15.2	−10.1	−5.1	0
$m_3 M_3$	0	0	0	−5.1	−10.1	−15.2	−20.2
Final Moments	0	−43.1	−120.2	−20.3	79.8	Symmetric	

The discontinuities produced at the Releases 2 and 3 by the external load applied to the
cut-back structure are

$$\theta_2 = \theta_3 = \int Mm_2 \, ds/\text{EI}$$

which by Simpson's rule

$$= 20(0 - 22 - 20 + 0 + 100 + 0 + 0)/3$$

$$= 3480/9$$

The discontinuity produced at Release 2 by the application of unit value of M_2 in (ii) is

$$f_{22} = \int m_2^2 \, ds/\text{EI}$$

which by Simpson's rule

$$= 20(0 + 4/12 + 2/10 + 12/16 + 2/4 + 4/48 + 0)/3$$

$$= 112/9$$

$$= f_{33}$$

where f_{33} is the discontinuity produced at Release 3 by the application of unit value of M_2 in
(iii).

The discontinuity produced at Release 2 by the application of unit value of M_2 in (iii) is

$f_{23} = \int m_2 m_3 \, ds/EI$

$= 20(0 + 4/16 + 2/4 + 4/16 + 0)/3$

$= 60/9$

$= f_{32}$

where f_{32} is the discontinuity produced at Release 3 by the application of unit value of M_2 in (ii).

Since there are no discontinuities in the original structure at the positions of the releases

$$\begin{bmatrix} f_{22} & f_{23} \\ f_{32} & f_{33} \end{bmatrix} \begin{bmatrix} M_2 \\ M_3 \end{bmatrix} = \pm \begin{bmatrix} \theta_2 \\ \theta_3 \end{bmatrix}$$

Due to symmetry,

$M_2 = M_3$

$\theta_2 = \theta_3$

$f_{22} = f_{33}$

$f_{23} = f_{32}$

Hence, $M_2 = M_3 = -\theta_2/(f_{22} + f_{23})$

$= -3480/(112 + 60)$

$= -20.2 \text{ kip feet}$

The final distribution of moments due to the prestressing force and secondary effects is given by

$M = Pe + M_2 m_2 + M_3 m_3$

and is tabulated in Table 3-7.

3-8. Fig. 3-8 shows details of a two span continuous bridge superstructure. The bridge is designed for two traffic lanes with HS15 live load. The superstructure dead load, including the weight of the $8^3/4 \times$ glued laminated Douglas Fir 24F-V4 stringers, is 60 pounds per square foot. The top and bottom faces of the stringers are fully supported laterally.

Requirements:

Based on stress consideration only, determine the required depth of the $8^3/4 \times$ glued laminated Douglas Fir 24F-V4 stringers.

Solution

The maximum positive moment in Span 12, produced by one lane of HS20 loading, occurs at Point 4 a distance of 10.7 feet from Support 1. Standard truck loading governs and this value is obtained directly from AISC Tables[3] as

$M = 231.4 \text{ kip feet.}$

Hence, the maximum moment produced by one lane of HS15 truck loading is

$M_{T4} = 0.75M$

$= 0.75 \times 231.4$

$= 173.6 \text{ kip feet}$

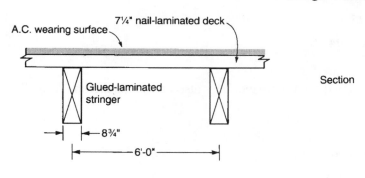

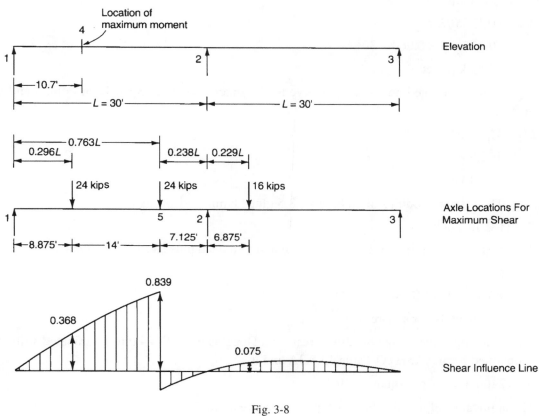

Fig. 3-8

Similarly, the maximum negative moment at Support 2 is given by

$M_{T2} = 0.75 \times 193.1$

$= 144.8$ kip feet

In accordance with AASHTO Section 3.8, impact need not be considered in timber structures. Hence, the maximum live load moments per stringer, after lateral distribution in accordance with AASHTO Section 3.23 for a stringer spacing of less than 6.5 feet, are given by

$M_{L4} = 0.50 M_{T4} S/4.25$

$= 0.50 \times 173.6 \times 6/4.25$

$= 122.5$ kip feet

$M_{L2} = 0.50 M_{T2} S / 4.25$

$\qquad = 0.50 \times 144.8 \times 6/4.25$

$\qquad = 102.2$ kip feet

The maximum dead load moments may also be obtained from AISC Tables[3] and are given by

$M_{D4} = 0.0703 SwL^2$

$\qquad = 0.0703 \times 6 \times 0.060 \times 30^2$

$\qquad = 22.8$ kip feet

$M_{D2} = 0.125 SwL^2$

$\qquad = 0.125 \times 6 \times 0.060 \times 30^2$

$\qquad = 40.5$ kip feet

Hence, the maximum dead plus live load moment occurs in Span 12 at Point 4 and is given by

$M_4 = M_{L4} + M_{D4}$

$\qquad = 122.5 + 22.8$

$\qquad = 145.3$ kip feet

The section modulus of an 8.75×28.5 glued laminated stringer is

$S = 1185$ inches^3

The stress produced in the stringer by the maximum dead plus live load is given by

$f_b = M_4 / S$

$\qquad = 145.3 \times 12{,}000/1185$

$\qquad = 1471$ pounds per square inch

The basic allowable design bending stress for a Douglas Fir 24F-V4 glued laminated stringer is obtained from AASHTO Table 13.5.3A and is

$F_b = 2400$ pounds per square inch

The applicable adjustment factors for bending stress are:

(i) $\quad C_v =$ volume factor given by AASHTO Equation (13-5)

$\qquad = (1291.5/bdL)^{1/10}$

$\qquad = (1291.5/8.75 \times 28.5 \times 22.5)^{1/10}$

$\qquad = 0.86$

where $L =$ distance between points of zero movement for a uniformely distributed load

$\qquad = 0.75 \times 30$

$\qquad = 22.5$ feet

(ii) $\quad C_M =$ wet use factor

$\qquad = 0.80 \ldots$ from Table 13.5.3A

The top and bottom faces of the stringers are fully supported laterally and lateral instability does not have to be considered.

Hence, the adjusted bending stress is

$$F'_b = C_v C_M F_b$$

$$= 0.86 \times 0.80 \times 2400$$

$$= 0.69 \times 2400$$

$$= 1656 \text{ pounds per square inch}$$

$$> f_b \dots \text{satisfactory}$$

In accordance with AASHTO Section 13.6.5, the stringers must be designed for the maximum shear occurring at a distance from the support given by the minimum of:

(i) $0.25L = 0.25 \times 30$

$$= 7.5 \text{ feet}$$

(ii) $3d = 3 \times 28.5/12$

$$= 7.125 \text{ feet} \dots \text{governs}$$

The influence line for shear at Point 5, located a distance of 7.125 feet from Support 2, is obtained from AISC Tables[3] and is shown in Figure 3-8.

The shear at Point 5 due to the standard HS15 truck is

$$V_{T5} = 24(0.368 + 0.839) + 16 \times 0.075$$

$$= 30.17 \text{ kips}$$

The undistributed shear due to one line of wheel loads is

$$V_{U5} = V_{T5}/2$$

$$= 30.17/2$$

$$= 15.08 \text{ kips}$$

The distributed shear due to the standard HS15 truck loading laterally distributed as specified for moment is given by

$$V_{S5} = 0.50 V_{T5} S/4.25$$

$$= 0.50 \times 30.17 \times 6/4.25$$

$$= 21.30 \text{ kips}$$

In accordance with AASHTO Section 13.6.5 the design live load shear force is

$$V_{L5} = (0.60 V_{U5} + V_{S5})/2$$

$$= (0.60 \times 15.08 + 21.30)/2$$

$$= 15.17 \text{ kips}$$

The dead load shear at Point 5 is obtained from the AISC Tables[3] as

$$V_{D5} = SwL(0.625 - 0.238)$$

$$= 6 \times 0.060 \times 30 \times 0.387$$

$$= 4.18 \text{ kips}$$

Hence, the maximum dead plus live load shear at Point 5 is given by

$$V_5 = V_{L5} + V_{D5}$$

$= 15.17 + 4.18$

$= 19.35$ kips

The shear stress produced in a stringer by this shear force is given by AASHTO Equation (13-9) as

$f_v = 1.5V_5/bd$

$= 1.5 \times 19.33 \times 1000/(8.75 \times 28.5)$

$= 116$ pounds per square inch

The basic allowable design shear stress for a Douglas Fir 24F-V4 glued laminated stringer is obtained from AASHTO Table 13.5.3A and is

$F_v = 165$ pounds per square inch

The applicable adjustment factor for shear stress is

C_M = wet use factor

$= 0.875 \ldots$ from Table 13.5.3A

The adjusted shear stress is, then,

$F = C_M F_v$

$= 0.875 \times 165$

$= 144$ pounds per square inch

$> f_v \ldots$ satisfactory

For a structure designed for HS15 loading, AASHTO Section 3.5.1 requires a design check using load combination Group 1A which is

Group $1A = \gamma(\beta_D D + \beta_L L)$

$= 1.0(1 \times D + 2 \times L)$

where, D = dead load

L = live load due to a single HS15 truck

and an overstress of 50 percent is allowed.

For bending moment, Group 1A is not critical.

For shear, the undistributed shear due to one line of wheel loads is

$V'_{U5} = V_{T5}/2$

$= 30.17/2$

$= 15.08$ kips

The distributed shear due to a single HS15 truck is

$V'_{S5} = 0.50 \times 0.50V_{T5}S/4.25$

$= 0.50 \times 0.50 \times 30.17 \times 6/4.25$

$= 10.65$ kips

The design live load shear force for a single HS15 truck is

$$V'_{L5} = (0.60 \, V'_{U5} + V'_{S5})/2$$

$$= (0.60 \times 15.08 + 10.65)/2$$

$$= 9.85 \text{ kips}$$

The total design shear force is, then,

$$V'_5 = V_{D5} + 2 \, V'_{L5}$$

$$= 4.16 + 2 \times 9.85$$

$$= 23.86 \text{ kips}$$

The shear stress produced in a stringer by this shear force is given by AASHTO Equation (13-9) as

$$f'_v = 1.5 \, V'_5/bd$$

$$= 1.5 \times 23.86 \times 1000/(8.75 \times 28.5)$$

$$= 143.5 \text{ pounds per square inch}$$

$$< 1.50 \, F'_v \ldots \text{ satisfactory}$$

Hence, the 8.75×28.5 Douglas Fir glued laminated stringers are adequate.

References

1. American Association of State Highway and Transportation Officials. *Standard Specifications for Highway Bridges*, Sixteenth Edition, as amended by the 1997 interim revisions. Washington, D.C. 1996.

2. Williams, A. *The Analysis of Indeterminate Structures*. Macmillan, London, 1967.

3. American Institute of Steel Construction. *Moments, Shears and Reactions: Continuous Highway Bridge Tables*. Chicago, IL, 1959.

4. Jenkins, W. M. Influence Line Computations for Structures with Members of varying Flexural Rigidity Using the Electronic Digital Computer. *Structural Engineer*. Volume 39, September 1961.

5. Wang, C. K. Matrix Analysis of Statically Indeterminate Trusses. *Proceedings American Society of Civil Engineers*. Volume 85 (ST4), April 1959.

6. Portland Cement Association. *Influence Lines Drawn as Deflection Curves*. Skokie IL, 1948.

7. Thadani, B. N. Distribution of Deformation Method for the Construction of influence Lines. *Civil Engineering and Public Works Review*. Volume 51, June 1956.

8. Lee, S. L. and Patel, P. C. The Bar-Chain Method of Analyzing Truss Deformations. *Proceedings American Society of Civil Engineers*. Volume 86 (ST3), May 1960.

9. Williams, A. The Determination of Influence Lines for Bridge Decks Monolithic with their Piers. *Structural Engineer*. Volume 42, May 1964.

10. Morice, P. B. and Little, G. *The Analysis of Right Bridge Decks Subjected to Abnormal Loading*. Cement and Concrete Association, London, 1956.

11. West, R. *Recommendations on the Use of Grillage Analysis for Slab and Pseudo-Alab Bridge Decks*. Cement and Concrete Association, London, 1973.

12. Loo, Y. C. and Cusens, A. R. A Refined Finite Strip Method for the Analysis of Orthotropic plates. *Proceedings Institution of Civil Engineers*. Volume 48, January 1971.

13. Davis, J. D., Somerville, I. J. and Zienkiewicz, O. C. Analysis of Various Types of Bridges by the Finite Element Method. *Proceedings of the Conference on Developments in Bridge Design and Construction, Cardiff, March 1971*. Crosby Lockwood, London, 1972.

14. Westergaard, H. M. Computation of Stresses in Bridge Slabs Due to Wheel Loads. *Public Roads*. March, 1930.

15. Portland Cement Association. *Handbook of Frame Constants*. Skokie, IL, 1948.

16. Lee, S. L. The Conjugate Frame Method and its Application in the Elastic and Plastic Theory of Structures. *Journal Franklin Institute*. Volume 266, September 1958.

17. American Association of State Highway and Transportation Officials. *Standard Specifications for Highway Bridges. Sixteenth Edition: Division I-A Seismic Design*. Washington, D.C. 1996.

18. Building Science Safety Council. *NEHRP Recommended Provisions for the Development of Seismic Regulations for New Buildings: Part 2, Commentary*. Washington, D.C. 1997.

19. Paz, M. *Structural Dynamics*. Van Nostrand Reinhold, New York, 1991.

20. Federal Highway Administration. *Seismic Design and Retrofit Manual for Highway Bridges*. Washington, D.C. 1987.

21. Hewlett Packard Company. *HP-48G Calculator Reference Manual*. Corvallis, OR 1994.

22. Portland Cement Association. *Notes on ACI 318-95: Building Code Requirements for Reinforced Concrete*. Skokie, IL, 1995.

23. American Concrete Institute. *Building Code Requirements and Commentary for Reinforced Concrete (ACI 318-95)*. Detroit, MI, 1996.

24. American Concrete Institute. *Analysis and Design of Reinforced Concrete Bridge Structures (ACI 343R-88)*. Detroit, MI, 1988.

25. Reynolds, C. E. and Steedman, J. C. *Reinforced Concrete Designers Handbook*. Cement and Concrete Association, London, 1981.

26. Kajfasz, S. Somerville, G. and Rowe, R. E. *An Investigation of the Behavior of Composite Beams*. Cement and Concrete Association, London, 1963.

27. Freyermuth, C. L. *Design of Continuous Highway Bridges with Precast, Pre-stressed Concrete Girders.* Portland Cement Association. Skokie, IL, 1969.

28. Freyermuth, C. L. and Shoolbred, R. A. *Post-Tensioned, Prestressed Concrete.* Portland Cement Association. Skokie, IL, 1967.

29. The Concrete Society. *Post-Tensioned Flat-Slab Design Handbook.* London, 1984.

30. American Association of State Highway and Transportation Officials. *Guide Specification for Alternate Load Factor Design Procedures for Steel Beam Bridges Using Braced Compact Sections.* Washington, D.C. 1986.

31. Nash, G.F.J. *Steel Bridge Design Guide: Composite Universal Beam Simply Supported Span.* Constructional Steel Research and Development Organization. Croydon, 1984.

32. Knowles, P. R. *Simply Supported Composite Plate Girder Highway Bridge.* Constructional Steel Research and Development Organization. Croydon, 1976.

Index Volume 3